T0135688

Bibliografische Information Der Deutschen Bibliothek

Die Deutsche Bibliothek verzeichnet diese Publikation in der Deutschen
Nationalbibliografie; detaillierte bibliografische Daten sind im Internet über
http://dnb.ddb.de abrufbar.

ISBN 3-8325-1307-8

Logos Verlag Berlin
Comeniushof, Gubener Str. 47,
10243 Berlin
Tel.: +49 030 42 85 10 90
Fax: +49 030 42 85 10 92
INTERNET: http://www.logos-verlag.de

Visualisation of biochemical pathways and their simulation results

Inauguraldissertation
zur Erlangung des akademischen Grades
eines Doktors der Naturwissenschaften
der Universität Mannheim

vorgelegt von

Dipl. Inf. (FH) Katja Wegner
Schriesheim

Mannheim, 2006

Dekan: Professor Dr. Matthias Krause, Universität Mannheim
Referent: PD Dr. Jürgen Hesser, Universität Mannheim
Korreferent: Professor Dr. Ing. Dr. h.c. Andreas Reuter, EML Research, Heidelberg

Tag der mündlichen Prüfung: 17. Februar 2006

In memory of my parents.

Contents

Contents

Contents

List of Figures

List of Figures

List of Tables

Danksagung

Diese Arbeit hat viel Zeit, Mühe und am Ende auch viele Entbehrungen gekostet. Deshalb danke ich meiner gesamten Familie und allen Freunden und Bekannten, die an mich geglaubt, mich in allen Dingen unterstützt haben und für die ich meist viel zu wenig Zeit hatte.

Insbesondere danke ich meinem Lebenspartner Andre Wengler, der am meisten zurückstecken und meine Launen ertragen musste. Trotzdem hat er mich immer wieder aufgebaut, ermutigt weiterzumachen und hat mich köstlich bekocht, so dass ich gestärkt an die Arbeit gehen konnte.

Ich danke Dr. Ursula Kummer und Dr. Ursula Rost für die Betreuung meiner Dissertation, die ebenso wie Ralph Gauges, Dr. Isabel Rojas, Dr. Sven Sahle und Jürgen Pahle am EML Research in Heidelberg für alle meine Fragen, ob fachlich, technisch oder privat, immer zur Verfügung standen und meine Arbeit voll und ganz unterstützt haben.

Desweiteren möchte ich mich bei Prof. Dr. Gerold Baier and Prof. Dr. Markus Müller für ihre gute Zusammenarbeit und ihre nette Gastfreundschaft in Mexico bedanken.

Ich danke PD Dr. Jürgen Hesser und Professor Dr. Ing. Dr. h.c. Andreas Reuter für ihre Anregungen und die Bereitschaft meine Dissertation zu begutachten.

Für das Korrekturlesen und für die guten Vorschläge beim Schreiben dieser Arbeit danke ich Dr. Ursula Kummer, Margot Mieskes, Dr. Ursula Rost, Dr. Isabel Rojas und Arun Natarajan.

Außerdem möchte ich mich beim gesamten EML für das super Arbeitsklima und bei der Klaus Tschira Stiftung für die finanzielle Unterstützung bedanken.

Ich danke Dr. Falk Schreiber für die Beantwortung meiner Fragen und die Bereitstellung von Screenshots und ich danke Michael W. Davidson für die Erlaubnis sein Zell-Bild im Einleitungsteil verwenden zu dürfen.

Abstract

Since researchers find out more and more about processes in living cells, modelling and simulation techniques are used to get new insights into these complex processes. The processes consist of reactions in which molecules are produced or rebuilt. Successive reactions build reaction pathways. These biochemical pathways can be modelled in mathematical equations which are solved by computers. This process is called simulation.

This thesis describes the Java package ViPaSi that supports the modelling process and the analysis of simulation results of biochemical pathways. It consists of three modules.

The module PathWiz offers a graph editor which allows the user to create a graph representation of a pathway. In this graph nodes represent molecules, while edges represent reactions. This representation helps researchers finding relationships between species, thus facilitating the modelling process. The nodes in these graphs should be placed dynamically according to biological conventions and a small number of edge crossings. Therefore, a new dynamic layout algorithm for biochemical pathways was developed and integrated in the PathWiz module. Furthermore, I supported the development of an exchange format for such layouts.

The other two modules support the visualisation of simulation results. The module SimWiz visualises multivariate data as a sequence of temporal snapshots with shape- or colour-coding on the basis of the graph representation. The module SimWiz3D shows the graph representation in the xy-plane and a tube for each node along the z-axis. These tubes represent the time series date for a certain time range and their diameters are increasing and decreasing according to concentration or number of particle changes. In addition, a number of user interactions and correlation methods for the analysis of the time series data are integrated, including a newly developed correlation method (matrix $\mathbf{A}$) for spatio-temporal data.

Zusammenfassung

Wissenschaftler finden immer mehr über "biochemische" Prozesse in lebenden Zellen heraus, damit wächst auch die Erkenntnis, dass die Prozesse äußerst komplex sind. Daher werden immer mehr Modellierungs- und Simulationsmethoden eingesetzt, die schneller Einblicke und neue Erkenntnisse liefern. Biochemische Prozesse bestehen aus Reaktionen, in denen Moleküle hergestellt oder verändert werden. Aufeinanderfolgende Reaktionen bilden Reaktionspfade. Diese biochemischen Pfade können als mathematische Gleichungen modelliert werden, die die Konzentrationsänderungen der Moleküle über die Zeit beschreiben. Diese Gleichungen werden von Computern gelöst und simulieren so den Ablauf dieser Pfade in der Zelle.

Diese Dissertation beschreibt das Java-Programmpaket ViPaSi, das die Modellierung solcher Pfade und die Analyse der Simulationsergebnisse unterstützt. Es besteht aus drei Modulen. Das erste Modul PathWiz bietet eine graphische Oberfläche zum Erstellen und Verändern von Graphen. Diese Graphen enthalten Knoten und Kanten, die Moleküle und Reaktionen darstellen. Diese Visualisierung hilft Forschern visuell Beziehungen zwischen den Molekülen zu erkennen und erleichtert somit den Modellierungsprozess. Die Knoten in diesen Graphen sollen automatisch entsprechend biochemischen Konventionen platziert werden, deshalb wurde ein neuer Algorithmus speziell für biochemische Pfade entwickelt und in PathWiz integriert. Außerdem wurde ein Format entwickelt, dass den Austausch solcher Graph-Visualisierungen mit verschiedenen Programmen ermöglicht.

Das zweite und dritte Modul visualisieren Simulationsergebnisse. Das Modul SimWiz repräsentiert multivariate Daten schrittweise, d.h. pro Bild wird genau ein Zeitschritt dargestellt und durch Farb- oder Größenänderungen in der graphischen Darstellung werden die Konzentrationsänderungen verdeutlicht. Im Gegensatz zu diesem Modul zeigt das Modul SimWiz3D die Zeitreihen im Ganzen zusammen mit der graphischen Darstellung. Jede Zeitreihe wird durch einen Schlauch in der dritten Dimension dargestellt, der entsprechend der Konzentrationsänderungen dicker oder dünner wird. Weiterhin wurden in dieses Modul viele Interaktionen und Korrelationsmethoden integriert, die die Analyse der gezeigten Daten unterstützen. Darunter befindet sich auch die neue Korrelationsmethode (A-Matrix), die speziell für Simulationsdaten entwickelt wurde, die aus vereinzelten Peaks bestehen, aber keine Phase haben.

1. Introduction and Motivation

"The purpose of computing is insight not numbers."
(R. Hamming (1973))

This dissertation is part of the interdisciplinary field of bioinformatics. But what does bioinformatics mean? The word bioinformatics consists of biology and informatics. Biology [4] can be defined as the science of life which studies the characteristics and the behaviour of organisms. Informatics or information science [4] is a branch of computer science which deals with structuring, creating, managing, storing, retrieving and transferring information. With these two definitions in mind bioinformatics can be defined as a science that tries to analyse biological facts by using mathematical and computational methods. The biological data is stored, organised and summarised in databases. Existing and newly developed algorithms implemented in computational programs are used to analyse this data. These analysing processes are supported by image processing and visualisation.

The analysed biological facts in this thesis deal with biochemistry which can be considered as the chemistry of life [4]. This biological field studies the smallest functional component each organism consists of - the living cell. Since cells have a nearly unlimited variety of forms and functions, a lot of information was gained by various experiments in laboratories. Thus, researchers discovered more and more facts about species and processes in living cells during the last years. The information is very complex and not yet complete. The following analogy [5] illustrates these problems.

> *Only few years ago ... colour TV sets were made*
> *up of thousands of separate components: transform-*
> *ers, transistors, condensers, resistors, coils, diodes,*
> *... and many other things. Let's assume an electron-*
> *ics specialist wants to reconstruct such an applicance,*
> *based on a complete parts list containing every single*
> *component with its precisely specified characteristics.*
> *Would he be able to join all the parts together to form*
> *a properly functioning system ...?*
> *Certainly not! ... without knowing the underlying cir-*
> *cuit diagram that shows how the various components*
> *have to be wired and connected with each other ...*

Researchers in systems biology have the same problem. They know the species in the cells but not all interactions and dependencies between them. Takahashi et al. [6] formulated this problem in the following questions:

How will we assemble the various pieces?

Moreover, they mentioned that the behaviour of a biochemical system will not longer be intuitive if the number of interactive species exceeds three. Therefore, computational and mathematical methods are used to supplement this knowledge and understanding of living cells. The numerical results of these methods need to be analysed and interpreted. For these reasons, Dr. Ursula Rost and myself developed the software package **ViPaSi** (**Vi**sualisation of biochemical **Pa**thways and **Si**mulation results) for the visualisation and analysis of biochemical networks and their simulation results.

The processes in the cell consist of single steps that produce, rebuild or reduce species in the cell. These steps are called reactions. Successive reactions form pathways (or networks). These pathways are often depicted as graphs. Such graphs contain nodes representing the participants (species) of the reactions and edges representing reactions and their directions. Since biological knowledge is changing very fast these graphs need to be drawn dynamically to provide flexibility in the context of different data and data updates. Conventional layout algorithms are not sufficient for every kind of pathway in biochemical research. This is mainly due to certain conventions to which biochemists/biologists are used to and which are not in accordance with conventional layout algorithms. Therefore, the first main goal of this thesis was the development of a new dynamic layout algorithm for graph visualisation of these reaction networks because this graph representation gives a first impression of the relationships between different species. These relationships build the basis for modelling and simulation processes.

Since experiments in laboratories become more and more complex as well as time consuming and expensive, modelling and simulating techniques are used. The biochemical reactions are commonly modelled as differential equations which describe concentration or particle changes of species over time. Solving such equations (mostly using computers) is called simulation. These simulations produce a lot of numerical data (time series). Therefore, a good visualisation of this data is needed, which was the second goal of this dissertation. I will discuss existing visualisation techniques and introduce a new visualisation tool that visualises time series in different views and provides a lot of user interactions to ease the exploration and interpretation of simulation data.

On the one hand a good visualisation is necessary and on the other analysis tools for this mass of data. Existing visualisation tools show a lack of interaction and analysis techniques. Therefore, I will introduce user interactions and correlation analysis methods that are integrated in the new visualisation tool. The most important question is: *how are the relations between the species in the network.* Mostly, correlation analysis is used to answer this question. Therefore, a new correlation method was developed which was the third main goal of this dissertation.

1.1. Outline

In chapter 2 the biological background, modelling, and simulation techniques will be described in more detail.

In chapter 3 the package **ViPaSi** and its three modules (**PathWiz, SimWiz, SimWiz3D**) will be introduced in general.

Chapter 4 will illustrate graph drawing fundamentals, existing algorithms and the new layout algorithm for the graph visualisation of biochemical pathways (module **PathWiz**).

3

In chapter 5 existing visualisation techniques for multivariate, time-dependent data and biochemical simulation results (e.g. **SimWiz**) will be introduced. Also, the module **SimWiz3D** for the a continuous time series visualisation will be described.

In the following chapter 6 the conventional cross-correlation matrix **C** and the new developed correlation matrix **A** will be explained.

Finally, in chapter 7 I will summarise the complete work.

2. Background

"A picture is worth a thousand words"
(Chinese proverb)

As I already mentioned in the introduction, biochemistry deals with studies of processes in living cells. For that reason, a short introduction to this field will be form the first section. Furthermore, modelling and simulation methods and visualisation techniques for time series data of existing tools are shortly described.

2.1. Biochemical processes

Biochemistry is the chemistry of life [4] that deals with studies of the smallest unit each organism consists of - the living cell. A cell can be compared to a production factory in which species are produced, reduced or rebuilt in single steps. Such single steps are called chemical reactions. These reactions are controlled by proteins that are similar to machines in a factory. The assembly lines of a factory are represented by successive reactions which build complete reaction pathways or networks. The schema of an animal cell can be found in figure 2.1 (plant cells are a little bit different). As seen in this figure the cell is not homogeneous but consists of different parts (called compartments), e.g. cell nucleus, cell membrane, endoplasmic reticulum and many more. Each compartment contains species which take part in reactions.

The more we find out about processes in living cells, the more complex the relationships between species in these processes will be. Therefore, modelling, simulation, and visualisation techniques are used to support the understanding of these processes. Furthermore, these techniques will reduce time and costs of experiments in laboratories and also the number of animal experiments in the case of drug development [5] because it is faster and easier to analyse a model under

5

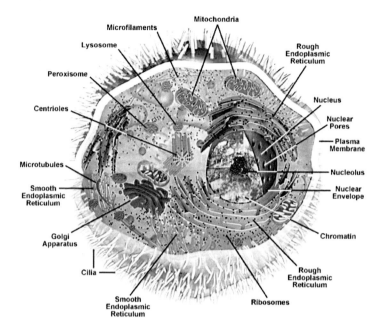

Figure 2.1.: The animal cell [1].

different conditions on the computer rather than experimenting in laboratories [7]. Moreover, models are applied to find small perturbations that cause large effects and can not be detected in experiments [7]. Consider as example, modelling and simulating is used in drug discovery to search for target species that can be used in therapy. Such computational methods reduce cost and time.

Typically, the strategy to analyse biochemical processes/systems can be divided into the following five steps [8]:

1. *making laboratory experiments*

2. *visualising biochemical pathways*

3. *modelling biochemical pathways*

4. *simulating and model analysis (e.g. parameter estimation or optimisation)*

5. plotting and visual interpretation of simulation results.

According to the first step, during experiments researchers find out the fundamental data for modelling and simulating biochemical systems. The visualisation of this knowledge in the second step helps researchers to identify relationships which ease the modelling process in the third step. The resulting model is simulated (fourth step) and the simulation results need to be interpreted to get new insights which is done in the last step.

The last four steps will be introduced in more detail in the following sections.

2.2. Biochemical pathways

A biochemical pathway is part of a complete biological system. Such a system has four main characteristics [2] (see table 2.1).

Structure	Which components take part and how are their structural relationships?
Dynamics	How does a system behave under different conditions?
Control	Which mechanisms control the behaviour of the system?
Design	Which rules built a biological system?

Table 2.1.: The four main characteristics of a biological system [2].

The first characteristic means that a biochemical pathway consists of species and their relationships. These relationships arise due to reactions between these species. The species of a reaction (called compounds) are separated into substrates and products. Substrates are used to build products. Furthermore, reactions can be catalysed by enzymes (special proteins) to run faster. Chemical reaction equations are used to describe these steps in a formal way (see figure 2.2).

The reaction arrow states the reaction direction. The substrates are placed on the left side of the arrow (the start point) and the arrow points to the products

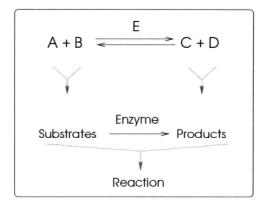

Figure 2.2.: An abstract chemical reaction equation. Arrows maintain the reaction direction and therefore which compounds are the substrates and which the products. Reactions are reversible which means they react forward and backward (" ⇌ ").

on the right side. In general reactions are reversible which is shown by two arrows in opposite directions (" ⇌ ") in the reaction equation. If the backward reaction is much slower than the forward reaction, the backward reaction can be ignored and the reaction equation is shown with a single arrow (" → "). Mostly, reactions in a pathway are successive which means the products of the first reaction are used in a second one as substrates. In other words the second reaction follows the first one.

Definition 1 *A* **biochemical pathway** *or* **network** *is defined as a list of compounds and successive reactions (a small example is shown in table 2.2).*

The visualisation of a pathway (second step in the analyses of biochemical systems in section 2.1) offers the advantage that the topology of the network which is tightly linked to its function, is easily depicted. In general a graph representation is used because the topology information is lost when a researcher is confronted with just a list of biochemical reactions. Voit [8] defined the following rules for constructing a proper graph:

1. list all participants

compounds	reaction equations
CO_2	$CO_2 + NH_4{}^+ - > Carbamolphosphate$
NH_4	$Carbamolphosphate + Ornithine - > Citrulline$
$Carbamolphosphate$	$Citrulline + Aspartate - > Argininosuccinate$
$Ornithine$	$Argininosuccinate - > Arginine + Fumarate$
$Citrulline$	$Arginine - > Urea + Ornithine$
$Argininosuccinate$	$Fumarate - > Malate$
$Arginine$	$Malate - > Oxalacetate$
$Fumarate$	$Oxalacetate - > Aspartate$
$Urea$	
$Malate$	
$Oxalacetate$	
$Aspartate$	

Table 2.2.: Small pathway example: Reactions of the urea cycle and a small part of the citrate cycle.

2. list all interactions

3. arrange all participants and interactions in the form of a graph.

Definition 2 A **graph** $G = (N, E)$ *visualises the relationships between objects. It consists of nodes and edges.* **Nodes** *(N) represent the objects and* **edges** *(E) the relationships between these objects [3].*

In the case of biochemical pathways nodes in the graph represent compounds and edges show which compounds are transformed into others (reaction) [8]. The edge direction symbolises the direction of the respective reaction. A graph representation of the example in table 2.2 can be found in figure 2.3.

Brandenburg et al. [9] differentiate between two types of pathway visualisation: static and dynamic visualisation. The static visualisation is mostly used in biochemistry books [10, 11], but examples can also be found on the computer, e.g. KEGG [12]. Being static, it offers no flexibility in the level of detail or in the exact information depicted by the respective graphs. To get rid of these problems dynamic visualisation techniques arose, that enable the user on the fly to see and

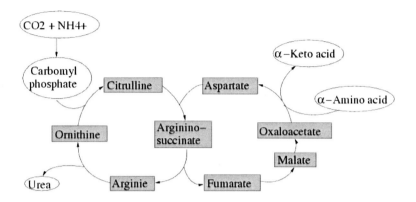

Figure 2.3.: The reactions of the urea cycle and a small part of the citrate cycle visualised as graph (see the reaction equations in table 2.2).

change only those pathways that are needed at the time of viewing [9]. For this reason, the development of a new dynamic layout algorithm especially for biochemical pathways is part of this dissertation and will be introduced in chapter 4.

The visualisation of compounds and their relationships can be used to find out design principles (fourth characteristic in table 2.1). Moreover, the visualisation gives an impression of the complexity of the system [8]. Dynamics and control mechanism (second and third characteristic in table 2.1) are studied by modelling and simulation tools which are the step three and four in analysing biochemical systems (see section 2.1) and will be described in the next section.

2.3. Modelling and Simulating biochemical pathways

Creating a graphical representation of a pathway is an important step to find out which compounds should be included into the modelling process. The modelling process translates a graph map into a mathematical structure [8] (third step in the analysis of biochemical systems introduced in section 2.1).

Definition 3 *A* **model** *is a mathematical description of the real world which predicts what the system is doing at any time step [13] and fulfils the following constraints [8]:*

- *captures realistic properties of the system (e.g. same reactions on inputs as the real system),*

- *is characterised by measurable parameters,*

- *simulation results must be comparable to real results, and*

- *the used mathematical description should be solvable by existing mathematical methods.*

Usually a model contains abstractions that are easier to handle as the real system [7]. That means the model is an approximation and can never predict the real processes precisely [14]. A model can be used to make hypotheses about the characteristics and the behaviour of a system, then these hypotheses can be verified by experiments in laboratories. These activities influence each other and improve understanding.

In the case of biochemical systems the model describes commonly concentration changes of all compounds over time. Differential equations represent these changes mathematically. These equations express *"the rate of productions of a component of the system as a function of the concentrations of the other components"* [15]. Unfortunately, this deterministic approach is not suitable for all kinds of biochemical systems. Therefore, stochastic methods will be used if the concentration values are very small and *"fluctuations in the timing of events"* occur [15]. These methods calculate the number of particles of the compounds instead of concentration changes. However, since more and more systems involve both deterministic and stochastic behaviour [16], hybrid methods are developed. They separate automatically the reactions in the model into deterministic and stochastic calculations [17].

After creating the model the fourth step in analysing a biochemical system in section 2.1 is the simulation. Simulation means that the equations of the model are evaluated with the computer by using different parameter settings [8]. It computes the dynamic behaviour of the underlying biological system [7].

The numerical simulation of deterministic equations approximates the concentration changes at fixed time points [15]. The stochastic approach evaluates probabilities for the occurrence of each reaction and the decision which reaction is next is based on chance. By interpreting the resulting time series data, researchers can get insights into dynamics and control properties of a system (table 2.1).

This was a really short introduction into modelling and simulating. To describe these techniques in detail is beyond the scope of this thesis since this thesis describes a new software package that supports the analysis of the results, but does not describe how to improve or change modelling and simulation techniques. Voit [8] gives deeper insights into this field.

The last step in the analysing schema for biochemical systems in section 2.1 is plotting and visualising simulation results. The next section will enter into this step by introducing visualisation techniques in existing modelling and simulation tools.

2.3.1. Visualisation of simulation results

In this section I will concentrate on the visualisation of simulation results in modelling and simulation tools according to step five in analysing biochemical systems in section 2.1. The standard visualisation technique is a concentration-time plot. The stated software tools in table 2.3 are free available and have the following minimal requirements:

1. simulation of any biochemical reaction network

2. visualisation of simulation results

3. graphical user interface (GUI)

As seen in table 2.3 all tools have integrated the general concentration-time plot visualisation but these plots are not suitable for a larger number of compounds (see figure 2.4).

Since the knowledge about the processes in the cell is increasing, the number of compounds in the modelled pathways is increasing, too. Therefore, new visualisation techniques are necessary. Only three tools mentioned in table 2.3 have further visualisation techniques, but they are mostly suitable for special purposes

Tool	Deterministic	Stochastic	Concentration-time plot	Other visualisations techniques	Platform
A-Cell [18, 19]	X		X		
BioNetS [20, 21]		X	X		
Cellware [22, 23]	X	X	X		
Copasi [24]	X	X	X		
DBSolve [25]	X	X	X		Windows
Dizzy [26, 27]		X	X		
Dynetica [28, 29]	X	X	X		
Ecell [30, 31]	X	X	X		Windows
Genesis [32]	X	X	X		
JigCell [33, 34]	X	X	X		
JSim [35, 36]	X		X		
JWS Online [37]			X		web-based
MesoRD [38, 39]		X	X	X	Windows
MetaboLogica [40, 41]	X		X		Windows
Moleculizer [42, 43]		X	X		
SBW [44, 45]	X	X	X		
SimFit [46, 47]	X	X	X		Windows
Simpathica [48]	X		X		
SmartCell [49, 50]		X	X		
Stochastirator [51]		X	X		GUI-Windows
StochSim [52]		X	X	X	GUI-Windows
Vcell [53]	X		X		web-based
XPPAUT [54]	X	X	X	X	

Table 2.3.: This table summarises the properties of existing modelling and simulation tools.

.

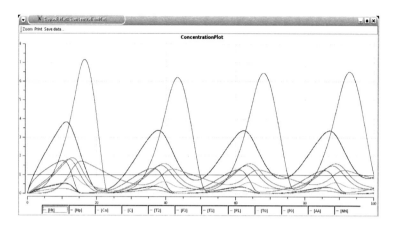

Figure 2.4.: Concentration-time plot of twelve species.

only, e.g. in MesoRD the visualisation of diffusion processes. None of them offers a dynamic layout algorithm to create the graph representation of a pathway automatically.

To summarize

In [55] I found a nice summarisation of this background chapter: "The simulation nature" (see figure 2.5). It summaries the single steps in the modelling and simulation area and the horizontal bars represent the amount of data passing between successive processes. Furthermore, it shows the necessarity of visualisation techniques which help researchers to handle, analyse and interpret the numerical simulation results.

For these reasons Dr. Ursula Rost and me developed the software package **ViPaSi**. It is a stand alone application that consists of three single usable modules. One module offers the possibility to create and layout dynamically biochemical pathways and the two other modules visualise simulation results in different ways (no concentration-time plots). This package can be used in addition to any modelling and simulation tool mentioned in table 2.3. In the next chapters **ViPaSi** and its modules will be introduced in detail.

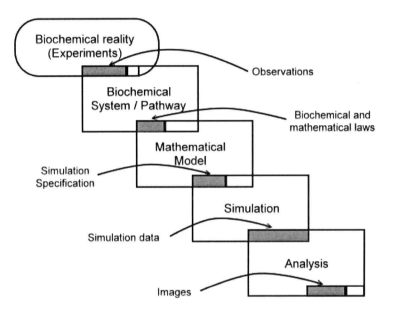

Figure 2.5.: "The simulation nature" [55]. The bars represent the different processes and the grey bares the amount of information that is handed over between the processes.

2. Background

3. ViPaSi

This chapter will give an overview about the developed package ViPaSi. It describes shortly the three modules of ViPaSi and the data exchange in this package.

3.1. Overview

ViPaSi is the abbreviation of **Vi**sualisation of biochemical **Pa**thways and their **Si**mulation results. The package is the result of this thesis and was developed to support the analysis of biochemical systems. It is a platform independent software package written in Java that can be used as add-on for any modelling and simulation software tool, e.g. the tools mentioned in section 2.3.1. It consists of three modules:

- **PathWiz** - graphical visualisation of biochemical pathways,

- **SimWiz** - visualisation of simulation results step-by-step,

- **SimWiz3D** - visualisation of simulation results in a continuous manner using three dimensions.

PathWiz supports the second step in the analysis of biochemical systems (see section 2.1), because it enables the user to create manually or dynamically graphical representations of biochemical pathways. The other two modules offer the possibility to visualise and analyse simulation results according to the fifth step in the analysis of biochemical systems in section 2.1.

Since it is a joint project of Dr. Ursula Rost and myself, figure 3.1 shows an overview of the package and marks parts implemented by Dr. Ursula Rost (UR) or myself (KW). Parts written by Dr. Ursula Rost are italicised in all figures in this thesis. As seen in figure 3.1 the PathWiz module is done by Dr. Ursula Rost and myself together. SimWiz was completely implemented by Dr. Ursula Rost and SimWiz3D by myself.

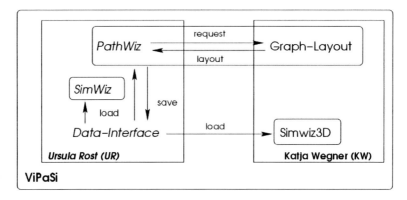

Figure 3.1.: Overview of ViPaSi showing the data exchange between the parts done by Dr. Ursula Rost (italicised) and by myself.

The overall package consists of 184 classes and nearly 49 000 lines of code of which 54 classes and 20 500 lines of code were implemented by myself.

3.2. Data exchange

SBML (Systems Biology Markup Language) [56] is used to exchange information about biochemical systems. SBML is a computer-readable file format to store biochemical reaction model information. It describes biochemical models used for simulations, e.g. participating reactions, their compounds and velocities. Over 90 tools already support this file format [57].

Since the graph representation of a biochemical pathway supports the modelling and analysing process, the layout information should also be saved into the SBML file. Therefore, Ralph Gauges, Dr. Sven Sahle, Dr. Ursula Rost and myself [58] developed together the first specification for a SBML layout extension. We discussed it in several SBML forum meetings with many SBML users and adapted it to further demands. Now it is accepted by the consortium and will be part of SBML level 3 (current version lever 2 [59]). This extension defines the storage of layout information in SBML files, e.g. coordinates and size of compounds as well as information about edge types. This extension does not include any rendering information, e.g. colour or shape properties, but it offers the possibility to store several layouts of the same model in one SBML file.

Ralph Gauges implemented our new layout specification as extension to the lib-SBML. The libSBML is a library for reading, writing, manipulating, translating, and validating SBML files and data streams [60]. It is not a complete application, but it can be integrated into applications to import and export SBML files in several programming languages [60].

ViPaSi uses the libSBML to read and write SBML information. For the internal communication an interface implemented by Dr. Ursula Rost is used to share model and layout information. This interface consists of the following three classes (see also the class diagram in figure 3.2):

- **GraphicalNode** - describes the properties of a compound and its position in the graph representation.

- **GraphicalEdge** - describes relationships between defined compounds.

- **GraphicalReaction** - collection of GraphicalNodes and GraphicalEdges which represent compounds and their relationships in one biochemical reaction.

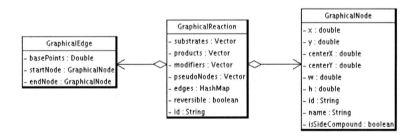

Figure 3.2.: Class diagram of the interface classes developed by Dr. Ursula Rost.

The relationships between the modules and this interface are shown in figure 3.3. The libSBML reads the model information from the file. This model information is stored in the interface classes. These classes are used to exchange this information between the modules. If the model information is changed or a graph layout was created, these changes are written into the SBML file by the libSBML.

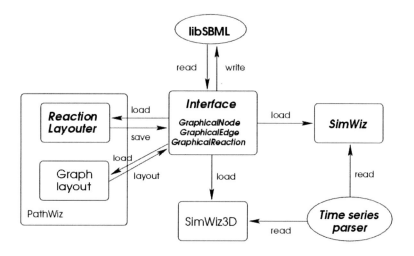

Figure 3.3.: Data exchange in ViPaSi.

The second exchange data format considers the simulation results. They must be saved in an ASCII file with a table structure. The first column corresponds to the time and all other columns to the participating compounds. The order of the compounds is not important, but the first or the second line of this table must contain compound names or SBML ids because a clear mapping between compounds in the used SBML model and compounds of the simulation results must be possible. Many of the simulation tools mentioned in section 2.3.1 support this ASCII format (example in table 3.1). The line that contains the number of steps in the data file is optional. The data values can be separated by a space or a tab.

By using the interface for the data exchange the modules are completely independent and changes in one module do not influence the functionality of other modules.

# 8							
# time	NADH	O2	PER3+	PER2+	O2-	COI	COII
0	0	0	843220	0	0	0	0
1.92239	116025	83009	843220	0	0	0	0
3.8412	231509	164609	843220	0	0	0	0
5.74499	346657	244640	843220	0	0	0	0
7.64555	461089	323846	843220	0	0	0	0
9.53534	575161	401654	843219	0	0	1	0
11.4144	688747	478230	843218	0	0	2	0
13.2855	801981	553454	843215	0	0	5	0
15.1523	914032	628233	843209	0	0	11	0

Table 3.1.: Example of an ASCII file containing simulation results. The first line contains the number of calculated time steps (without the initialisation step) and the second contains the species names or SBML ids. The line containing the number of time steps is optional.

3.3. PathWiz

PathWiz is a joint module by Dr. Ursula Rost (UR) and myself (KW). It consists of two submodules:

1. **ReactionLayouter** (UR) - visualises reaction pathways

2. **GraphLayout** (KW) - creates a dynamic layout of a pathway

3.3.1. ReactionLayouter (UR)

The ReactionLayouter [61] is a simple graph editor with a GUI implemented by Dr. Ursula Rost. On one hand the user can create his own graph from scratch or edit existing graph layouts by moving, inserting, deleting nodes or edges, redefining font or node sizes. On the other the user can load network information from a file. The list of reactions comprising the biochemical network can be submitted as a SBML or as a simple text file (listing all reactions separated by semicolons). It is visualised by a hyper-graph, which means that each reaction is represented by

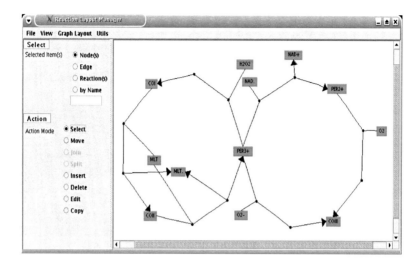

Figure 3.4.: Screen shot of PathWiz with an example pathway layout of a small part of the PO reaction. Reactions are represented by two connected dummy nodes. The first dummy node is connected to the substrates and the second one to the products of the respective reaction.

two connected dummy nodes, one is linked with all substrates and the other one with all products of this reaction.

Often there is a differentiation between two types of compounds, main and side compounds. Main compounds lie on the backbone of the pathway, e.g. in linear pathways they participate in successive reactions [62]. All other compounds in this pathway are considered as side compounds. Therefore, nodes coloured in grey represent side compounds and nodes coloured in turquoise represent main compounds. This differentiation of compounds is used in the graph layout process. How this works will be explained in detail in chapter 4.

Since a manual layout is very time-consuming [61], pathways can be layouted by using the second submodule: GraphLayout. At the end the layout information can be saved in a SBML or SVG file.

3.3.2. GraphLayout (KW)

This submodule consists of classes to create a layout for a given biochemical pathway. The description and development of the used graph layout algorithm will be part of chapter 4 and is the first main goal of this thesis.

With the aid of the libSBML this module can also be used as an command-line tool which reads an SBML file and writes the resulting layout information into this file. Then, this file can be visualised in any SBML viewer like JDesigner[63] or the ReactionLayouter submodule of ViPaSi. Therefore, it can also easily be added into any GUI or database application.

3.4. SimWiz (UR)

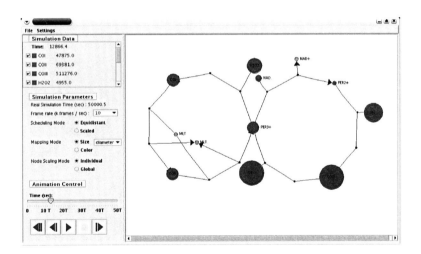

Figure 3.5.: Screen shot of SimWiz. It displays a time step of a small part of the PO pathway [64] where the diameter of the circles corresponds to the simulation values of the compounds at the current time step.

SimWiz [61] is one of two modules in this package which visualises simulation results. As you can see in figure 3.5 the GUI contains a visualisation and a control panel. The visualisation is based on the graph representation given in a SBML file and the time series data is mapped onto this graph by modifying the shape or the colour of the nodes. Each image shows one time step. Either the colour or the diameter of the circle representing a node corresponds directly to the concentration value or number of particles of the according compound at a certain time step.

In the animation mode a sequence of such images is created and is shown as a video in which either the colour changes or the node size increases or decreases automatically. With the aid of the control panel the user can decide which compounds should be visualise and how fast the animation should be.

3.5. SimWiz3D (KW)

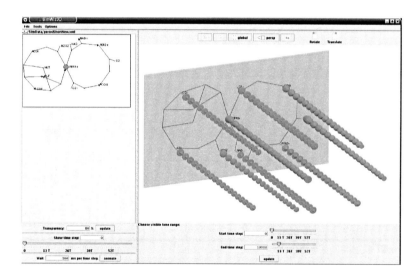

Figure 3.6.: Screen shot of SimWiz3D, shows a part of the PO pathway. The tubes represent the time series data of the corresponding node for a certain time range.

SimWiz3D is the second visualisation module of ViPaSi. It visualises time series data using a three-dimensional representation and provides the user with a lot of different ways to interact with the simulation data and utilise it. In contrast to SimWiz it visualises complete time series in a single 3D image and not only one single time step per image. Furthermore, it offers a lot of user interactions and analysis methods that are not available in SimWiz. Nevertheless, the 2D visualisation of single time steps in SimWiz was adapted because it helps users to locate single time steps in the 3D representation and compare concentration values. A detailed description of this module can be found in chapter 5.

3.6. The software architecture

Each module of ViPaSi is a stand-alone application that is based on the model-view-controller concept (MVC) [65]. The MVC concept separates the user interface from the programme logic to isolate the functionality from GUI changes. This feature gives the programmer the flexibility to rebuild the GUI without touching the functionality and vice versa. The application will be divided into a model, a view and a controller part (see figure 3.7).

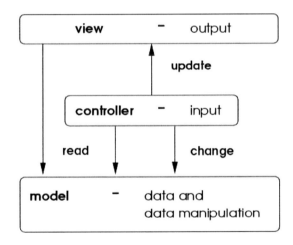

Figure 3.7.: The MVC architecture.

The model part stores data and encapsulates the core functionality of the application that makes the model independent from the view and the controller. The view and the controller together form the user interface of the module (output +

input). The model is state free and does not know how or how often its data is visualised.

The view has access to the data in the model and presents it to the user, but it does not change the data. It has only a read-only representation of the model.

The controller handles user inputs and translates them into requests on the model or view. It calls methods to change or manipulate the data in the model but does not copy data values to the view. Since I implemented a state free model, the controller is responsible for sending update requests to the view that reads data changes and redraws its display.

The view and the controller are strongly related because the output and the input of an application are strongly related. The view depends on the model since if the model is changed, the view must be updated, too. The controller invokes model changes and has to inform the view about these changes.

I chose this architecture because it makes the modules very flexible in different ways. Firstly, every view can be changed or substituted by another one without changing algorithms or data structures. Therefore, the package is more robust. Secondly, one model object allows that multiple simultaneous views (e.g. SimWiz3D) can be active at the same time. Thirdly, they are synchronised since they are representing the same model state. Fourthly, by splitting the package into model, view and controller, it can easily be extended, maintained and reused, e.g. the step-by-step visualisation of the SimWiz package is reused in SimWiz3D. Each part has its own responsibilities and the interface in figure 3.2 or events are used to communicate between them. Finally, the partitioning allows parallel implementation which was necessary for the PathWiz module that is a joined module.

According to these requirements I also reviewed the Presentation-Abstract-Control (PAC), the micro kernel, and the reflection architecture, but the MVC concept matched best. A description how this architecture was realized in the ViPaSi package will be part of the implementations sections of the following chapters.

4. Visualising biochemical pathways

> *"Graph drawing is the best possible field I can think of:*
> *It merges aesthetics, mathematical beauty and wonderful algorithms.*
> *It therefore provides a harmonic balance between the*
> *left and right brain parts."*
>
> *(D.E. Knuth)*

Biochemical reaction pathways are often depicted as graphs. With the aid of such graphs researchers find faster the relationships between participating species than by using a textual list of these reactions. At the beginning graph representations were handmade, but with the use of modelling and simulation techniques the size of these graphs is increasing much faster. For this reason, these graphs should be drawn dynamically to provide flexibility in the context of different and changing data. Since standard graph layout algorithms are not suitable for all kinds of biochemical pathways, dynamic layout algorithms have been developed specially for these pathways. These algorithms have drawbacks in finding small cycles and cycles that share nodes.

For these reasons, this chapter will introduce a new dynamic graph layout algorithm for the graph representation of biochemical pathways which is part of the module PathWiz. Firstly, fundamental terms and algorithms from the graph drawing field will be described. Secondly, existing algorithms for biochemical pathways together with their advantages and disadvantages will be introduced. Finally, the new layout algorithm will be explained and compared to existing algorithms.

4.1. Background - Graph drawing

Graph drawing as part of graph theory can be found in many applications, e.g.

- databases (entity-relationship diagrams)

- information systems (organisation charts)

- software engineering (data flow diagrams)

In 1994 the bibliography of Battista et al. [3] already contains about 300 papers with a lot more applications. This was later continued by Ivan Herman [66]. These two examples show how extensively studied this field is. In the next section I will define the graph theory terms which are used in this dissertation. A more detailed introduction and most of the definitions can be found in [3] or [67].

4.1.1. Definitions

As defined already in chapter 2 (definition 2) a graph $G = (N, E)$ consists of a set of nodes N and a set of edges E.

Definition 4 *The* **edge** *$e = (u, v)$ connects node u and node v ($u, v \in N$). Therefore, u and v are the* **end nodes** *of this edge and* **adjacent** *to each other. Adjacent nodes are called* **neighbours**. *The* **degree** *of node v corresponds to the number of its neighbours.*

There is a differentiation between directed and undirected graphs.

Definition 5 *A* **directed graph** *(or* **Digraph***) consists of directed edges.*
A **directed edge** *$e = (u, v)$ points from node u to node v ($u, v \in N$). This edge is an* **outgoing** *edge of node u and an* **incoming** *edge of node v. Therefore, the number of incoming edges is called* **indegree** *and the number of outgoing edges* **outdegree**. *In addition, nodes without any outgoing edges are called* **sinks** *and without any incoming edges* **source**.

In contrast to these definitions an undirected graph does not contain any directed edges.

Definition 6 *A* **bipartite graph** *contains two node sets N_1 and N_2 ($G = (N_1, N_2, E)$) and only nodes from different sets are adjacent: $\forall\ e \in E$ with $e = (w, x)$, $w \in N_1$, $x \in N_2$.*

Furthermore, it is important to define what a path and a cycle is.

Definition 7 *A (directed)* **path** *is a chain of nodes $(n_1, n_2, n_3, \ldots, n_i)$, $n_i \in N$, such that e=$(n_i, n_{i+1}) \in E$ for $1 \le i \le k - 1$. A (directed) path is a (directed)* **cycle** *if $(n_k, n_1) \in E$ and does not contain any edge twice.*

Definition 8 *In a* **connected graph** *there is a path between each node pair $(u, v) \in N$.*

With the definition of a directed cycle an **acyclic graph** can be defined as a directed graph that does not contain any directed cycles. Below are the definitions of subgraphs and connected components.

Definition 9 *A* **subgraph** *$G' = (N', E')$ is part of a graph G when $N' \subseteq N$ and $E' \subseteq E \cap (N'xN')$.*

Definition 10 *A* **connected component** *of G is a maximal connected subgraph. Instead, a* **strongly connected component** *[68] has one more condition: each pair of nodes n_1 and n_2 in this subgraph must be neighbours. Therefore, both directed edges $e_1 = (u, v)$ and $e_2 = (v, u)$ must exist.*

The above mentioned definition build the fundamental graph terminology used in this dissertation. In the next section I will shortly introduce important graph algorithms.

4.1.2. Graph algorithms used

The algorithms in this section are used to find paths between nodes in a graph. A detailed explanation of these algorithms can be found in [68].

4. Visualising biochemical pathways

Depth first search:

The depth first search algorithm (short DFS) searches "deeper" in a graph whenever possible. Therefore firstly, the start node (step 1 in figure 4.1) will be expanded. Secondly, one successor (step 2) of the start node will be expanded and then one successor (step 3) of the successor of the start node and so on (see figure 4.1).

A classical application of the DFS is the search for (strongly) connected components in a graph.

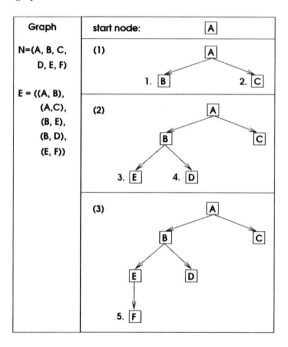

Figure 4.1.: An example of the depth first search algorithm.

Breadth first search:

In contrast to the DFS this algorithm searches in the breadth and finds the shortest instead of the longest path between two nodes. Therefore, firstly the breadth first search (short BFS) expands the start node (step 1 in figure 4.2). Secondly, all successors of the start node are expanded after each other (step 2 and 3 in figure 4.2). Thirdly, the successors of the successors are expanded.

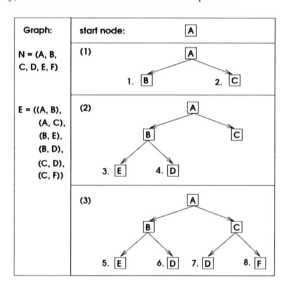

Figure 4.2.: An example of the breadth first search algorithm.

In the next section standard graph layout techniques will be introduced.

4.1.3. Graph layout

A **graph layout** is an algorithm that maps each node $n \in N$ to a distinct point X and each edge $e = (u, v) \in E$ to a curve with the end points X_1 and X_2 [3]. The usefulness of a resulted layout depends on the readability to recognise fast and clearly the shown information. The readability depends on aesthetic criteria, e.g. low number of edge crossings and bends. Some more criteria are listed in table 4.1. I separated these criteria in three groups: according to edges, bends and the layout in general. In addition to the first aesthetic criteria you can demand that the graph layout does not contain any edge crossings which is called a **planar layout**.

edge crossings:	minimising the total number of edge crossings
total edge length:	minimising the sum of all edge lengths
maximum edge length:	minimising the maximum length of an edge
uniform edge length:	minimising the variance of edge lengths
angular resolution:	maximising the smallest angle between two edges on the same node
total bends:	minimising the number of bends along an edge
maximum bends:	minimising the maximum number of bends along an edge
uniform bends:	minimising the variance of number of bends along an edge
area:	minimising the used area of the layout to e.g. save screen space
aspect ratio:	minimising the ratio between the width and the height of a the rectangle covering the layout
symmetry:	display symmetries of the graph in the layout

Table 4.1.: Aesthetic criteria for graph layouts [3].

Furthermore, you can draw up a list of constraints which the graph layout should fulfil. Some constraints are shown in table 4.2.

These criteria and constraints often conflict with each other because some of them exclude each other [3]. Moreover, a lot of criteria and constraints cannot be realized in an algorithm at the same time [67]. Therefore, most layout algorithms are only suitable for particular applications and solve an optimisation problem according to the given criteria. In the next sections standard layout algorithms that are used for drawing biochemical pathways are briefly described.

centre:	placing particular nodes close to the centre of the layout
external:	placing particular nodes on the outer boundary of the layout
cluster:	placing particular nodes close together (e.g. nodes of a subgraph)
left-right-sequence:	placing nodes of a path aligned from left to right
top-bottom-sequence:	placing nodes of a path aligned from top to bottom

Table 4.2.: Some layout constraints [3].

Hierarchical layout approach

The hierarchical layout algorithm produces a layered digraph. The nodes of the given graph are assigned to layers, such that connected nodes are placed into different layers from top to bottom (table 4.2 last row) according to the edge direction (see also figure 4.3).

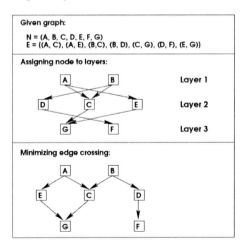

Figure 4.3.: Hierarchical layout algorithm, an abstract example.

The hierarchical layout creates a hierarchical ordering of nodes. If an edge $e =$

(u, v) points from node u to node v, node u will be assigned to layer L_1 and v to layer L_2. Edges between nodes in the same layer are not allowed.

The first idea of this algorithm was established by Warfield [69] and Carpano [70]. Sugiyama et al. [71] developed the most popular method which was extended by Eades and Sugiyama [72] and is used in many applications (e.g. efficient implementation in [73]). Two newer algorithms were invented by Healy et al. [74] (consider node height and width) and Sander [75] (hypergraph layout).

There are also edge crossing reduction algorithms (e.g. [76, 77, 78, 79], experimental study in [80]). These algorithms create an order of nodes in the layers, such that the edge crossings are minimised (table 4.1 first row).

Force-directed approach

The force-directed (or spring-embedder) algorithm simulates the graph as a physical system. Edges are springs with certain forces and nodes are charged particles. Nodes that are not connected repel each other. Iterative methods are used to reach a state of least total energy. This algorithm creates a highly symmetric layout which fulfils the last aesthetic criterion in table 4.1 (see also figure 4.4), but does not reduce edge crossings since only the edge length is taken into account by the forces.

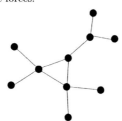

Figure 4.4.: An example layout.

Eades [81] invented the first force-directed algorithm. The performance of this algorithm can be improved by using the preprocessor developed by Muton et al. [82]. A more robust and faster computation was done by Fruchterman et al. [83].

Kamada et al. [84] improved the force-directed approach by using springs with different length and strength. In addition, there are many other heuristics, e.g.

1. Frick et al. [85] - a fast adaptive method for undirected graphs,

2. Gajer et al. [86] - a divide and conquer method for large graphs in 2D and 3D,

3. Chuang et al. [87] - using potential fields,

4. Hachual et al. [88] - using potential fields for large graphs,

5. Gansner et al. [89] - using a stress function instead of an energy function

Circular layout approach

The circular layout algorithm places as many nodes as possible on the circumference of a cycle. The diameter of the cycle depends on the number of nodes. Nodes are symmetrically put on the circumference with the same distance. Figure 4.5 shows an example of this algorithm. Mostly, a radial tree layout is used to realize a

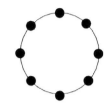

Figure 4.5.: An example layout.

circular layout algorithm (see [3], [90] - minimum number of circles). Six et al. [91, 92] developed another circular layout algorithm (groupings are user-defined) which was extended by Kaufmann [93]. Unfortunately, sometimes algorithms produce cycles in which neighbouring nodes on the circumference are not connected.

The divide and conquer approach

This approach is mostly used in graph drawing when subgraphs are important. Firstly, the graph is divided into subgraphs, e.g. according to their topology. Secondly, these subgraphs are layouted:

- cyclic subgraphs → circular layout

- linear subgraphs → hierarchical layout.

Finally, all subgraphs are joined to one complete graph, e.g. [94, 95].

Planar drawing

A graph is called planar if its layout in the 2D plane does not contain any edge crossings. Since edge crossings reduce the readability of a layout, planar drawing plays an important role in graph drawing [96, 97]. In a planar drawing the plane is partitioned into topologically connected regions called faces [3] and the edges of a node are ordered clockwise around it. Diestel [98] defines a planar graph by the following properties:

1. $N \in R^2$,

2. every edge is an arc between two nodes,

3. different edges have different sets of end nodes,

4. the interior of an edge contains no vertex and no point of any other edge.

Derived from Euler's formula the upper bound on the number of edges of a planar graph with at least three nodes is $3n - 6$ edges [99]. Furthermore, there are a lot of planarity testing algorithms. The first linear time algorithm was developed by Hopcroft et al. in 1974 [100]. This algorithm uses the following divide and conquer method [3]:

- A graph is planar if and only if all its connected components are planar.

- A connected graph is planar if and only if all its biconnected components are planar.

In this approach the graph is decomposed into connected and biconnected components. Therefore, the planarity test is reduced to testing the planarity of connected components [3]. Other algorithms use st-numbering [101, 102] and PQ trees [103, 104].

There are also strategies to make non-planar graphs planar. The most common planarization technique is edge deletion. A small number of edges is deleted from the graph to receive a planar graph [96]. This approach is equivalent to the method of finding the maximum planar subgraph. This problem is NP-hard and Tamassia et al. [105] developed an efficient solution. The resulting planar graph is layouted and then the deleted edges are reinserted such that the number of edge crossings is small [99]. This edge reinsertion corresponds to the crossing minimisation problem that is also NP-hard [106]. Crossing minimisation is mostly used when a hierarchical layout is created to find an appropriate order of nodes in the layers using node permutations. It will be NP-complete even if there are only two layers [106].

The second widely used technique constructs a graph layout with a small number of edge crossings and replaces each crossing and its two edges ($\{a, b\}, \{c, d\}$) by a new node z and four new edges ($\{a, z\}, \{z, b\}, \{c, z\}, \{z, d\}$) [99]. The resulting planar graph is then drawn by a planar drawing algorithms. Finally, the inserted nodes are resubmitted by crossings.

The third technique is called node splitting or vertex splitting [97]. It is used to create graphs with certain properties, e.g. planar graphs. A node v can be split into several nodes $v_1, ..., v_k$, such that these nodes cover the adjacencies of v in the original graph. In general, nodes are considered one at a time. A node is added to the graph if the resulting layout remains planar. Otherwise the node is split, such that the graph is planar. The determination of the minimum number of nodes that have to be split to planarize a graph is NP-complete [107].

I apply the 2-way splitting in the new layout algorithm that means that the node v is split into exactly two nodes v_1 and v_2. This splitting is used in two different cases. Firstly, nodes are split to find cycles that share nodes. If a node is part of a found cycle, it will be split. The first node gets the edges to nodes in this cycle and the second gets all remaining edges. Secondly, if the layout process is finished, nodes are split to reduce edge crossings or planarize the graph. If an edge has edge crossings, one end node of this edge will be split. The first node receives this edge and will be placed near to the other end node of this edge and the second node receives all other edges.

After introducing fundamental terminology and algorithms the graph representation of biochemical pathways will be described in more detail and these layout algorithms will be discussed according to the visualisation of biochemical pathways.

4.2. Biochemical pathways as graphs - State of the art

As I have mentioned in chapter 2, a list of reactions in a biochemical pathway is generally visualised as a graph. This representation is more informative and shows interactions in a clearer way than a textual list of reactions.

In the case of biochemical pathways nodes in the graph represent compounds and the edges the interactions between these compounds. The compounds in a reaction are divided into substrates and products. The substrates are consumed to build the products which corresponds to the direction of the reaction and therefore, to the direction of an edge. If an edge points from n_1 to n_2, n_1 is the substrate and n_2 the product of the respective reaction. Most existing visualisations use a bipartite graph representation in which one node set representing reactions is connected to the other node set representing compounds (see figure 4.6). The rendering of nodes and edges is very different which will be shown in the next sections when some drawings of existing projects will be introduced.

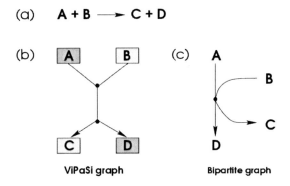

Figure 4.6.: This figure shows how the abstract reaction in (a) is visualised in ViPaSi (b) and as bipartite graph (c) in most existing tools.

In ViPaSi the compounds are represented by boxes containing the name of the compound as a label and edges are represented as straight lines. Furthermore, two connected dummy nodes are used to represent one reaction and enables the user to differentiate clearly between substrates and products (black dots in figure 4.6a). All substrates of the reaction are connected to the first dummy node and all products are connected to the second one. Thus, the graph contains two sets of nodes: nodes representing compounds and dummy nodes representing reactions. Nevertheless, this graph is not a bipartite graph because in bipartite graphs only edges between nodes of different node set are allowed. Since the two dummy nodes are connected by an edge this constraint is not fulfilled.

In ViPaSi the differentiation between side and main compounds is done by different colours. The main compounds have a turquoise (dark grey) box background and the side compounds a grey one. In most other tools the side compounds are set next to the edge between the main compounds of the respective reaction and are connected by curved lines to this edge.

As already mentioned in chapter 2 there is a differentiation between static and dynamic graph visualisation of biochemical pathways. Therefore, firstly static visualisation tools are introduced and secondly dynamic visualisation tools. A review containing visualisation tools for biological networks can be found in [108].

4.2.1. Tools with static pathway visualisation

In this section I will introduce three projects as representatives of the static visualisation of biochemical pathways. I chose the glycolysis pathway as an example. This pathway transforms glucose (sugar) into pyruvate and builds energy needed by any organism.

BioCarta

BioCarta [109] is a web-based tool which contains static drawings of pathways. Figure 4.7 shows the glycolysis pathway. Compounds are represented with their chemical form and beneath their names. Enzymes are shown only by numbers.

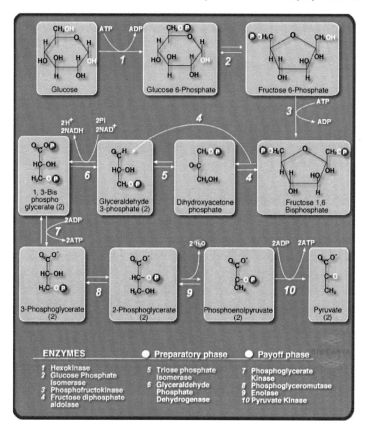

Figure 4.7.: Static pathway visualisation of glycolysis in BioCarta [109].

ExPASY

ExPASY Biochemical Pathways [110, 111] gives online access to the scanned-in version of the Biochemical Pathways wall chart by Michal [10] (figure 4.8). This chart is partitioned into pieces. A keyword search allows the user to find the interesting pieces. Figure 4.8 shows a part of the glycolysis pathway. Different colours are used to differentiate between compounds and enzymes. Furthermore, the colour of the reaction arrow indicates in which organism this reaction was observed.

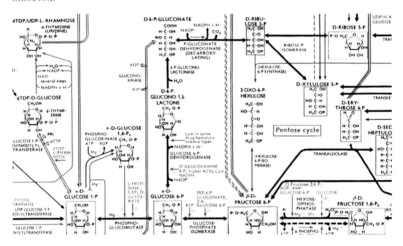

Figure 4.8.: A part of the glycolysis pathway in ExPASY [111] (static visualisation).

KEGG

KEGG (Kyoto Encyclopedia of Genes and Genomes) [112, 113] contains pathway maps which are in the "GIF" file format and continuously updated.

Figure 4.9 shows the glycolysis pathway. Compounds are presented by small circles with the their names next to this circle. Enzymes are visualised by rectangles with their distinct EC-number (standardised enzyme classification). The semi-rectangles point to adjacent pathways.

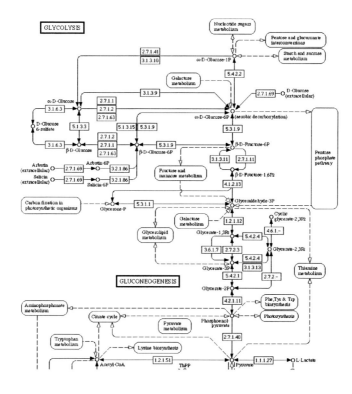

Figure 4.9.: Static pathway visualisation of glycolysis in KEGG [113].

To summarise

Although these static figures are suitable for the representation of biochemical pathways, there are some problems [9]. Firstly, they are made manually which is time consuming. Secondly, researchers have to decide in advance whether they want to draw an overview or details of the pathway. Thirdly, the finished drawing cannot be changed easily. In the end if the shown data or the user focus has been changed, the whole figure must be either edited or redrawn. For these reasons dynamic visualisation is used to compensate these problems. Tools using standard layout algorithms for dynamic visualisations of biochemical networks will be described in the next section.

4.2.2. Tools with dynamic visualisation using standard layout algorithms

Dynamic graph layout algorithms try to visualise the graph in such a way that it is easy to survey. This means that crossing of edges is avoided as much as possible. Nodes and labels have to be placed in such a way that they do not overlap. These tasks can be performed by standard graph layout algorithms. The following free available tools make use of standard layout algorithms to visualise pathways. Some tools offer different standard layout algorithms but for each pathway layout only one algorithm is usable (no combination of algorithms).

- **Bioconducter** [114]: uses the graph visualisation software Graphviz [115] (hierarchical, circular or force-directed layout algorithm)

- **BioMiner** [116]: hierarchical layout algorithm

- **Cytoscape** [117]: hierarchical, cyclic or force-directed layout algorithm (no combination possible)

- **Osprey** [118]: different circular layout algorithms

- **PathBinder** [119]: Graphviz

- **PathCase - Pathway database system** [120]: hierarchical or force-directed layout algorithm

- **Pathway Miner** [121]: Graphviz

- **Pathway Studio** [122]: force-directed layout algorithm

- **PATIKA** [123]: improved force-directed layout algorithm with directional and regional constraints

- **PaVESy** [124, 125]: hierarchical or force directed layout algorithm

- **VisAnt** [126]: force-directed layout algorithm

- **VitaPad** [127]: Graphviz

- **WebInterViewer** [128]: force-directed layout algorithm - for faster results: graph separated in connected components which are layouted individually and reassembled to one whole graph by the force-direct layout algorithm

- **yWays** [129]: hierarchical or force-directed layout algorithm

Short conclusions

Standard layout algorithms as used in all applications above are not sufficient for every kind of pathway in biochemical research. One reason is the very high degree of connectivity in complex biochemical networks, and the other reason is that there is a certain representation that biochemists are used to that does not match the way a standard layout algorithm would represent a complex biochemical network (see [130] for a detailed discussion). As an illustration of these problems see figure 4.10 which shows different visualisation of the pathway example in table 4.3. These reactions build one cycle of four nodes and seven nodes which are connected to this cycle. The fourth reaction contains compound B as substrate and as product.

A + B → C;
C → D + E + F;
F → D + E + B;
B → G + H + D + B;

Table 4.3.: Example of an abstract reaction pathway.

Since each standard layout algorithm is only suitable for a certain graph topology, the results are visually not satisfactory. The hierarchical layout creates a

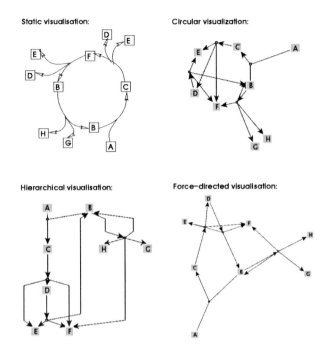

Figure 4.10.: Different visualisations of the abstract example in table 4.3. The single small black dots represent reactions.

hierarchical ordering which is not suitable for cyclic graphs as in the example (table 4.3). Therefore, it can only be used to layout complete hierarchical biochemical pathways.

The force-directed layout tries to create a symmetric drawing but you cannot recognise the cycle and it produces some edge crossings since it takes only the edge length into account.

The circular layout separates the graph in connected components and places nodes of the same component on a circumference of a cycle. Because of this strategy, on the one hand the resulting layout displays a cycle in which the neighbouring nodes on the circumference are not connected. For that reason, it is not a correct cycle according to the definition of a cycle in definition 7. On the other hand researchers only want to see circular structures if they represent correct cycles.

For these reasons, several specific dynamic layout algorithms for metabolic pathways have been developed in the past and are shortly described in the next section.

In addition, compound B is substrate and product in the fourth reaction. This fact will be best visualised if compound B is split into two nodes as seen in the handmade example. Since none of these classic layout algorithms and existing algorithms for metabolic pathways is able to split nodes, they would not be able to create a layout similar to the handmade example.

4.2.3. Layout algorithms developed for biochemical pathways

As already mentioned there is a differentiation between main and side compounds. Main compounds lie on the backbone of the pathway, e.g. in linear pathways they participate in adjacent reactions [95]. All other compounds in this pathway are considered as side compounds. This differentiation is important during the layout and rendering process of the following layout algorithms. In general, only main compounds are placed by the layout algorithms and side compounds are set near the edge representing the reaction in which they take part.

BioCyc

Karp et al. [95, 62] developed a divide-and-conquer layout algorithm for the project BioCyc [131]. In the first step the graph containing only main compounds is decomposed into subgraphs. These subgraphs are drawn accordingly to their topology (linear → hierarchical graph layout, cyclic → circular graph layout, branched → tree layout). In the second step a hierarchical layout algorithm assembles these subgraphs to a whole graph. The result for glycolysis is shown in figure 4.11.

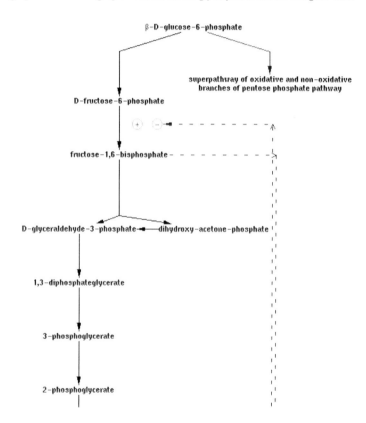

Figure 4.11.: Dynamic visualisation of a part of the Glycolysis in BioCyc.

Layout algorithm of Becker et al.

In 2001 Becker et al.[94] developed a divide-and-conquer method similar to Karp et al. Unlike Karp et al. they only distinguish between cyclic and hierarchical subgraphs in the graph and decompose found subgraphs with the help of a force-directed algorithm [132, 81]. However, this method is only able to handle main compounds.

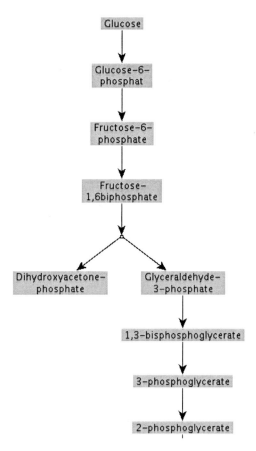

Figure 4.12.: A part of the Glycolysis pathway drawn by the layout algorithm of Becker et al..

PathDB

As part of the PathDB project Mendes et al. (personal communication) developed a PathwayViewer tool that consists of an improved Sugiyama (hierarchical layout [71]) and an individual circular layout algorithm to place the main compounds in a pathway. Additionally, they allow the user to edit the final drawing which is an advantage when an improvement of the resulted layout would be necessary. A screen shot of the application and the glycolysis pathway is shown in figure 4.13.

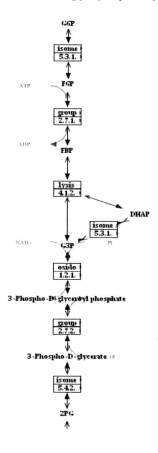

Figure 4.13.: Dynamic visualisation of a part of Glycolysis in PathDB.

49

BioPath

BioPath [133, 130] is a dynamic electronic version of the *Boehringer Biochemical Pathway* map by Michal [10, 11]. It uses an improved Sugiyama hierarchical layout algorithm [71]. However, BioPath is not longer available, thus I could not test its properties.

Layout algorithm of Rojdestvenski

In contrast to the other projects Rojdestvenski [134] uses a modified spring-embedding layout algorithm [84] for 3D-representations mainly. The algorithm considers main and side compounds as nodes during the layout process but with different priorities. Firstly, only the main compounds are placed. Secondly, the algorithm is started again with the main compounds and side compounds at the same time, but with frozen coordinates for the main compounds.

Short conclusions

The approaches BioCyc, BioPath and PathDB were implemented to visualise data in databases and make more or less use of additional information in their specific underlying databases, e.g. information about the order of reactions, side compounds etc. This is an advantage when aiming at the best graphical representation of data in the database, e.g. metabolic pathways. However, it often restricts the usage of the layout algorithms to these applications. Simulation and modelling tools that also want to need sophisticated visualisation techniques cannot rely on additional data in most cases. A graph layout algorithm used in these tools has to work, e.g. on the basis of the information as presented in a SBML file [56]. SBML files only contain explicit data about the individual reactions present in a specific biochemical network.

Furthermore, the approaches BioCyc, BioPath and PathDB first calculate the coordinates of the main compounds by the according layout algorithm and subsequently place the side compounds separately as labels near the edge representing the reaction in which they take part (one single label for each occurrence). In figure 4.13 you see that this side compound placement techniques leads to overlapping labels.

In contrast to these projects algorithm of Rojdestvenski treats side compounds as nodes of the graph and their coordinates are determined with the spring-embedder algorithm. Furthermore, each side compound occurs only once instead of one node for each occurrence in the network. However, this approach leads to many edge crossings in a graph with highly connected side compounds.

All of the above algorithms work reasonably well for small to medium sized networks. However, the complexity of studied biochemical networks is increasing. For large and complex pathways the existing pathway layout algorithms often face problems with respect to the number of edge crossings. Such complex pathways contain highly connected nodes and cycles that share nodes with other cycles. In addition, biological conventions stress the importance of cycles, even small cycles in general since such structures represent important recycling processes and short-cuts in the system. However, existing algorithms do not take these conventions into account. As an illustration of these problems see figure 4.14 which shows the visualisation of the Peroxidase-Oxidase reaction (PO reaction [64]) done by the algorithm of Becker et al.. These elementary reactions interact strongly with each other and two important recycling processes are present. The algorithm of Becker et al. created a tangle of nodes and edges which is impossible to survey.

In the following sections I will show how the new algorithm solves these problems, reduces edge crossings and represents the reactions in a way that corresponds more closely with biological conventions. Furthermore, the new algorithm is able to find cycles that share nodes which is not supported by any of the existing algorithms.

4.3. A new dynamic layout algorithm for biochemical pathways

Because of the problems mentioned in sections 4.2.2 and 4.2.3 with existing standard layout algorithms and algorithms for biochemical pathways, a new layout algorithm was developed which tackles these problems by reducing the number of edge crossings in complex systems and taking new conventions into account to identify and visualise cycles. This is achieved by identifying even small cycles and by splitting nodes to improve the readability of the dynamic drawing. Fur-

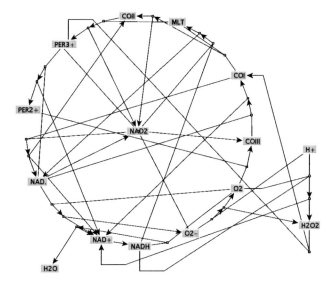

Figure 4.14.: Dynamic visualisation of the PO reaction [64]. This picture is a result of the algorithm of Becker et al.. The 17 reactions of this pathway are hardly recognisable.

thermore, the algorithm is independent from database information in order to be easily adopted in any application.

I have studied both static graphs of biochemical pathways and also how researchers create such graphs manually. I derived the following constraints and criteria that should be fulfilled by the new layout algorithm:

1. Consider biological conventions:

   a) separate the pathway in cyclic and hierarchical components,

   b) layout these components according to their topology,

   c) cycles look like cycles (all nodes on the circumference of a circle are connected),

   d) search for the smallest cycle instead of the longest one to emphasise recycling processes or short cuts in a pathway, and

   e) find cycles that join compounds.

2. Differentiate clearly between substrates and products

3. Minimise edge crossings

4. Minimise edge length and place nodes representing compounds in the same reaction near each other

5. Avoid label overlapping

6. Be independent of additional information

As a basis for this work of mine I have chosen the algorithm of Becker et al., because it is able to find cyclic and hierarchical structures that are the two basic topologies in which every complex biochemical pathway can be separated. This feature fulfils the above mentioned constraints 1a) and 1b). Furthermore, the algorithm of Becker et al. is not linked to a specific database that fulfils constraint 6). In addition, only two (Becker et al. and Karp et al.) of the five introduced layout algorithms for biochemical pathways are able to separate a graph into cyclic and hierarchical subgraphs and two (Becker et al. and Rojdestvenski) of five are not linked to a database.

4.3.1. The algorithm

As starting point for the new algorithm the implementation of the algorithm of Becker et al. is used. In general, the new algorithm differs from the algorithm of Becker et al. in the following ways: it is able to join and split nodes, and to detect smallest cycles or cycles of arbitrary size instead of just the longest one. For a better comparison see the sequence diagrams of both algorithms in appendix A. In addition, since the algorithm of Becker et al. is not able to handle side compounds, this feature was included.

Similar to Karp et al. [62] the definition of side compounds results from a predefined list with compound names (see table 4.4). That list is editable by the user. Each compound (side and main) is treated as a node in the graph. However, in contrast to Rojdestvenski [134] the new algorithm places side and main nodes simultaneously which means that as default side compounds have the same priority as the main compounds, e.g during the process of cycle search, main compounds are prioritised. This default is chosen, because many examples show that the differentiation between main and side compounds is helpful at times, but often somewhat arbitrary blurring the biochemical reality. Nevertheless, it is also possible to generate a layout without any side compounds.

ATP	DATP	ADP	DADP
AMP	CTP	CDP	GTP
GDP	ITP	IDP	UTP
UDP	CO2	FAD	FADH
FADH2	H2O2	NAD	NADH
NADP	NADPH	GLUTAMINE	GLUTAMATE
H2O	H+	2-OXOGLUTARATE	O2
O2-	NH3	NH4	NH+
NAD+	NAD2	NADP+	NAD.
MLT	MLT.	PI	PPI
SO4-			

Table 4.4.: Default list of side compounds used by the new algorithm.

In the following paragraphs, I will describe in detail how the new algorithm works.

Identifying subgraphs

This section describes the first part of the algorithm which identifies cyclic and hierarchical subgraphs of a given pathway. To find joined cyclic subgraphs nodes which are part of more than one cycle are split.

The pseudo-code in figure 4.15 describes the modified recursive method of Becker et al. to identify circular and hierarchical subgraphs in a given pathway. The first step is to search for the smallest instead of the longest cycle (figure 4.15, line 1). As explained above, this procedure is chosen since otherwise biologically relevant information might get lost, since small cycles often represent important recycling processes or short cuts in a pathway. One example is shown in figure 4.16 in which all graphs represent a part of the PO reaction [64]. The first picture was crafted with a graphic program by a biochemist. The second one was dynamically generated with the algorithm of Becker et al.. Here, the two cycles of the first picture are not easy to depict because of the emphasis on the longest possible cy-

Method:	findSubgraphs(*graph, subgraph*)
Input:	*graph* - all nodes of the pathway
	subgraph - selected nodes of the graph (first method call: *subgraph = graph*)
Output:	*foundCycles* - all found cyclic subgraphs of this pathway
	foundHiers - all found hierarchical subgraphs of this pathway
	foundSingleNodes - all single nodes which are not part of any found subgraph

```
 1.  cycle ← searchSmallestCycle(subgraph)
 2.  if (cycle.size() > 0) // cycle was found
 3.      if (cycle contains all nodes of subgraph) // one big cycle
 4.          doCircularLayout(cycle)
 5.          foundCycles.add(cycle)
 6.      else // complex graph
 7.          doCircularLayout(cycle)
 8.          foundCycles.add(cycle)
 9.          // Split all nodes in the cycle that have at least
10.          // two edges to nodes outside the cycle.
11.          subgraph ← splitCycleNodes(cycle, subgraph)
12.          // Create a list of all nodes of subgraph which are not part of cycle.
13.          remainingNodes ← getRemainingNodes(subgraph, cycle)
14.          components ← getConnectedComponents(remainingNodes)
15.          for (each nodeSet in components)
16.              findSubgraphs(graph, nodeSet)
17.  else // cycle is empty, no cycle was found, subgraph is hierarchical
18.      joinSplitNodes(subgraph)
19.      doHierarchicalLayout(subgraph)
20.      hierComponents ← getConnectedComponents(subgraph)
21.      for (each nodeSet in hierComponents) do
22.          if (number of nodes in nodeSet is 1)
23.              foundSingleNodes.add(nodeSet)
24.          else
25.              foundHiers.add(nodeSet)
```

Figure 4.15.: Pseudocode of method *findSubgraph()*: Find all hierarchical and cyclic subgraphs of a given pathway.

cle. However, the two cycles represent the two main recycling processes of enzyme intermediates and are therefore crucial for the reaction mechanism. These cycles are shown in the third picture in figure 4.16 which is generated by the new layout algorithm.

Definition 11 *The* **smallest cycle** *must consist of at least three compounds and must contain all dummy nodes of each reaction to which the compounds in this cycle belong.*

However, the minimum number of nodes in a cycle is adjustable by the user, since there are of course cases where the cycle representing biologically important information is not the absolutely smallest. Therefore, if the first layout depicting the smallest cycle is not of the desired quality, the user can change it by increasing the number of compounds for the cycle search.

In the first round of the cycle search only the main compounds and dummy nodes are used to find the smallest cycle. Then, the algorithm looks for connected components. The algorithm of Becker et al. looks for strongly connected components instead. In contrast to connected components, strongly connected components consider the edge direction. Thus, the path between each pair of nodes in the respective strongly connected components is only valid if the direction of the edges is always the same. By using connected components the algorithm is able to find both cycles where all edges have the same direction (e.g. in figure 4.16 the left cycle ($PER^{3+}-COII-COI$) in the first picture) and also cycles where edges have different directions (e.g in figure 4.16 the right cycle ($PER^{3+} - PER^{2+} - COIII$) in the first picture). Finally, the modified breadth first search (BFS) [68] finds the smallest cycle if existing.

If no cycle is found in the first call of the *findSubgraph* method, the whole graph will be drawn hierarchically. Otherwise, if one cycle has been found already and the method does not detect a second one with only main and dummy nodes, nodes representing side compounds are also included into the cycle search.

The algorithm keeps distinguishing between these three cases:

- No cycle found → draw the complete graph with a hierarchical layout algorithm (figure 4.15, line 17)

- All nodes of the graph belong to the found cycle → draw the complete graph with a circular layout algorithm (figure 4.15, line 3)

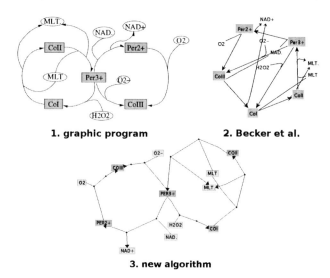

1. graphic program   **2. Becker et al.**

3. new algorithm

Figure 4.16.: Visualisation of a part of the PO reaction [64] (6 reactions). The layout of the graph as seen in the first picture was crafted with the aid of a graphic program by a biochemist, the second one was dynamically generated by the algorithm of Becker et al. and the third one by the new algorithm.

- Complex graph → draw the found cycle with a circular layout algorithm and separate the remaining nodes of the pathway into further cyclic and hierarchical subgraphs (figure 4.15, line 6)

In the first case the algorithm of Becker et al. uses a standard hierarchical layout algorithm. This standard algorithm was improved by separating the placement of the nodes into two steps. Firstly, the main and dummy nodes are placed by the standard algorithm. Secondly, the side compounds are split to create as many nodes as occurrences in reactions exist. Hence, every node has only one edge. These nodes are positioned one layer above or under the other end node of the respective edge according to the direction (top to down).

In the second case the whole graph consists of one cycle and all nodes are therefore positioned by a standard circular layout algorithm.

In the complex graph case the graph consists of various circular and hierarchical

4. Visualising biochemical pathways

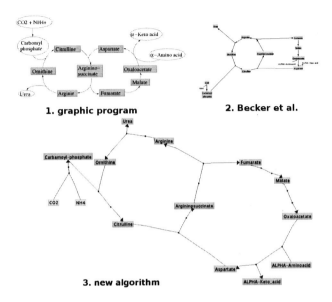

1. graphic program

2. Becker et al.

3. new algorithm

Figure 4.17.: Visualisation of the urea cycle and parts of the citrate cycle. The first picture was drawn manually, the second one was dynamically generated by the algorithm of Becker et al. and the third one by the new layout algorithm. All visualisations look similar but the two cycles in the first and third picture are not well represented in the second one.

subgraphs. Since in all existing layout algorithms each node is part of exactly one subgraph, these algorithms are not able to find cycles which share nodes. Therefore, I added the possibility to split (figure 4.15, line 11) and join nodes (figure 4.15, line 18 and figure 4.19, line 2). The first picture in figure 4.17 shows the urea cycle and a part of the citrate cycle crafted with a graphical program by a biochemist. These two cycles are connected at the compound **argininosuccinate**. The algorithm of Becker et al. finds the urea cycle and considers the unshared parts of the citrate cycle as hierarchical subgraph (second picture in figure 4.17). In contrast to this picture the new algorithm finds two cycles joined at **argininosuccinate** (third picture in figure 4.17) according to the biochemist's picture.

This result is achieved by splitting nodes in found cycles which could also be

58

part of another cycle (figure 4.15, line 11). These nodes must represent compounds and must have at least four edges, two edges to nodes in the found cycle and at least two edges to nodes which are not part of this cycle. Dummy nodes are not allowed to split.

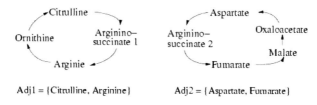

Adj1 = {Citrulline, Arginine}   Adj2 = {Aspartate, Fumarate}

Figure 4.18.: Splitting nodes. An example of how a node, in this case argininosuccinate, is split into two nodes. The adjacency lists (Adjx) show which edge belongs to which node.

Definition 12 Splitting nodes*: A node n is split into n_1 and n_2 when it is part of a cycle and has at least two edges to nodes that are not in this cycle. Node n_1 gets all edges connecting nodes in the found cycle and n_2 all remaining edges, see also figure 4.18.*

In this way, several cycles representing biochemical reaction cycles that share a compound can be found and represented accordingly.

When no more cycles are found, the remaining nodes are regarded as hierarchical. Split nodes are joined and the subgraph is inspected to find connected components before the improved hierarchical layout algorithm places the nodes of this subgraph (figure 4.15, line 18). Each connected component is considered as one hierarchical subgraph but components with only one node are saved in an extra set (figure 4.15, lines 20-24) and are placed separately. In contrast the algorithm of Becker et al. put all remaining nodes in one hierarchical subgraph. If the subgraph consists of different connected components, this technique could lead to unnecessary edge crossings when the subgraphs are reassembled to a complete graph.

Building the complete graph

After detecting all the subgraphs, they are then reassembled to a complete graph
by

- Joining split nodes if possible (= joining cycles at one node, figure 4.19, line
  2).

- Searching for further cycles in the found subgraphs (figure 4.19, lines 3-7).

- Reassembling the found subgraphs to a complete graph using a force-directed
  layout algorithm (figure 4.19, line 8).

- Reducing edge crossings between subgraphs (figure 4.19, lines 9-18).

For the joining of split nodes present in cycles, the algorithm tries to join main
nodes with priority over side nodes. Therefore, cycles with split main nodes will
be joined before cycles with split side nodes.

Definition 13 Joining two nodes: *The according cycles are rotated and moved*
together at these nodes. One node will be deleted from the graph and all its edges
will be shifted to the other one.

Only two cycles are allowed to be joined at the same node because more than
two cycles would cause edge crossings (see an example in figure 4.16 (PER^{3+})).
For the detection of further cycles, the algorithm searches for subgraphs which
are connected by at least two edges with another subgraph. These edges must have
different source and target nodes. If such edges exists, the algorithm will determine
the shortest path between the nodes of these two edges in both subgraphs. The
nodes found in these paths are used to build a new cycle. This new cycle must
also correspond to the above explained definition of a valid cycle and is then
drawn accordingly. The new cycle and the already existing one will be joined
at the common nodes. See an example in figure 4.17, the two cycles are joined
at argininosuccinate and the four dummy nodes of the two reactions in which
argininosuccinate participates.

In contrast to the algorithm of Becker et al. the final reassembling step starts after all subgraphs are found and split nodes are joined. The cycle with the maximal

number of edges to other subgraphs is the central subgraph. The force-directed algorithm places all subgraphs and single nodes around the central cycle to build the complete graph. The algorithm of Becker et al. uses the longest cycle (maximum number of nodes) as central subgraph.

Finally, to reduce edge crossings, all edges between found subgraphs are checked. The list of these edges is sorted by their length in descending order because typically the longer the edge the higher the number of edge crossings. Starting from the longest edge the number of edge crossings is counted for each edge of this list and nodes are split to reduce the number of crossings. This number of allowed edge crossings can also be changed by the user.

Method:	doLayout(*graph*)
Input:	*graph* - complete *graph*
Output:	*newGraph* - new drawing of *graph*

  1. *foundCycles, foundHiers, foundSingleNodes* ← findSubgraphs(graph)

  2. joinFoundCyclesOnSplitNodes(*foundCycles*)

  3. **// Search for further cycles which share at least two nodes.**

  4. **// If two subgraphs are connected by two edges, build a new cycle**

  5. **// with the nodes of these edges and the nodes connecting these edges.**

  6. **// Join both cycles at the common nodes.**

  7. searchForFurtherCycles(*foundCycles, foundHiers*)

  8. reassembleSubgraphs(*foundCycles, foundHiers, foundSingleNodes*)

  9. **// Reduce edge crossing: Count crossings for edges between subgraphs.**

10. **for** (*edges*)

11.   **check:** *noOfCrossings* ← countCrossings(*edge*)

12.   **if** (*noOfCrossings* > *allowedCrossings*)

13.   **if** (one of the nodes is split)

14.   moveEdgeOverToSplittingPartner()

15.   continue at **check**

16.   **else**

17.   splitTargetNode()

18.   moveNewNodeNearToSourceNode()

Figure 4.19.: Overview of the layout process.

Before splitting a node the algorithm checks whether one end node of this edge was the result of another splitting operation and checks whether its splitting partner could take over this edge with at most two edge crossings. If this is not possible, the target node of the examined edge is split and a new node is created that has only this edge and is placed near the source node. Nodes with only one edge are moved directly near the other edge end node and are not split. To generate a planar graph the number of allowed edge crossings can be set to zero but that also increases the number of split nodes.

In addition, the used standard layout algorithm of the YFiles graph library [135] (hierarchical, force-directed, circular layout) take node sizes into account which avoids label overlapping (constraint 5).

4.3.2. Discussion

I have presented the new dynamic layout algorithm for metabolic pathways. One of the main differences to the existing algorithms is the emphasis on finding small cycles. This results in a biochemically meaningful representation in many cases since cycles in biochemical networks often stand for important processes like recycling of intermediates, energy or electron carrier producing or futile cycles. Therefore, biochemists are used to seeing these processes as graphical cycles and the new algorithm takes care of this convention. For those cases where the smallest meaningful cycle does not match the default settings, they can be easily adjusted. In figure 4.20 you see the layout of the 17 reactions of the PO reaction done by the new layout algorithm. That figure shows a much better representation of this pathways than the algorithm of Becker et al. in figure 4.14.

The algorithm is able to handle linear, cyclic and complex metabolic pathways considering main and side compounds. A complex pathway consists of diverse hierarchical and cyclic subgraphs. Nodes are split and joined to improve the detection of these subgraphs and to minimise edge crossings. Finally in many cases the drawing reflects the biological context better than previous approaches, e.g. cycles sharing nodes can be found and represented. Therefore, the new algorithm satisfies the following constraints:

1. *Considering biological conventions:* The algorithm uses a divide and conquer method which separates the graph into circular and hierarchical subgraphs. This subgraphs are layouted by circular and hierarchical graph layout algorithms.

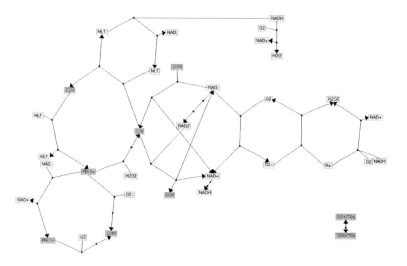

Figure 4.20.: Visualisation of the PO reaction [64]. The picture is a result of the new layout algorithm and shows the reactions in a much clearer way as compared to figure 4.14.

By identifying cycles sharing nodes and by splitting nodes, pathways are drawn more accurately and more similarly to the visualisations biologists are used to (e.g. figure 4.17).

The algorithm searches for small cycles instead of the longest one to emphasise e.g. recycling processes.

2. *Unequivocal distinction between substrates and products:* Each reaction is represented by two connected dummy nodes which allows an exact distinction between products and substrates in each reaction.

3. *Edge crossing minimisation:* The edge crossings are minimised in the two splitting phases. The first phase splits highly connected nodes which could be part of different cycles and the second one splits an end node of an edge with more than two edge intersections.

4. *Edge length reduction:* Both dummy nodes of one reaction must be part of

the same subgraph which minimises the distance between two dummy nodes. As mentioned above, compounds are split and placed near to each other to minimise edge crossings which also reduces the edge length. These methods also fulfil the constraint that compounds of the same reaction should be placed *near each other*.

5. *Avoid label overlapping:* Since the used standard layout algorithms consider node sizes, label overlapping can be avoided.

6. *Complexity reduction:* The complexity of pathways is reduced by splitting higher connected nodes (see also above). Thereby it is possible to untangle nets of many edges to one node, see KEGG [113] pyruvate metabolism (pathway no. 00630, e.g. Glyoxylate).

Since the new algorithm is based on the algorithm of Becker et al. , it calculates similar results for the examples optimally represented in the Becker at al. publication. Several examples drawn by the algorithm of Becker et al., the algorithm of Karp et al., and the new dynamic layout algorithm can be found in section 4.4. The Mendes et al. algorithm (PathDB) can only be used by a general user in the context of the respective database and therefore just with the pathways stored in those.

In the case of the Karp et al. (BioCyc) algorithm which is also normally used in the context of a database, I was able to compare the algorithm in an isolated manner, since it was generously supplied by Karp and coworkers (in section 4.4.2). The isolated algorithm performed well on small to medium sized samples, however, faced some problems w.r.t. edge crossings when considering larger or higher connected pathways. In addition to the information on the individual reactions, the isolated algorithm also uses information about the order of the reaction events which is absent in model files, e.g. SBML files.

The Mendes et al. algorithm and to a lesser extend the Karp et al. algorithm usually use additional information about the considered pathway from their database, e.g. the order of reaction events as pointed out above, to simplify the layout process.

Since such information is not available to a simulation/modelling tool, the new algorithm relies solely on a list of reactions of the pathway and optionally a predefined list of side compounds. The existing layout algorithms for metabolic pathways treat the side and main compounds of a pathway differently from the new algorithm. They all treat the side compounds as labels, which results in the labels overlapping in complex pathways. Although side compounds are part of the graph in the new algorithm the algorithm produces similar results compared to the existing algorithms in those cases where the latter produce good results and solves the overlapping problem in more complex cases.

The user defined list of side compounds naturally influences the treatment of the nodes during the layout process. When editing this list the user should keep in mind that different side compound lists lead to different drawings of the same pathway, see figures 4.21 and 4.22.

Since some parts of the algorithm rely on stochastic methods, the same pathway could be represented with different layouts. For example, the force-directed layout algorithm could cause different assembling of subgraphs. In addition the breadth first search of the cycle finding process checks the nodes according to their number of edges in descending order. If there are different nodes with the same number, the order is chosen randomly which means a varying order of nodes could lead to different cycles.

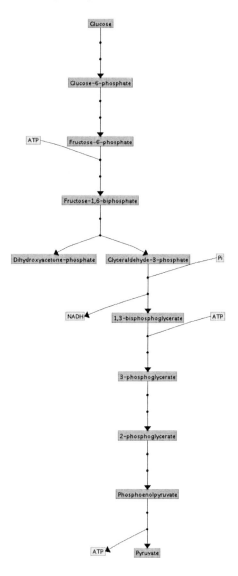

Figure 4.21.: Visualisation of glycolysis (1) done by the new layout algorithm. In this picture ATP, Pi and $NADH$ are considered as side compounds which displays the known hierarchical structure of the glycolysis pathway.

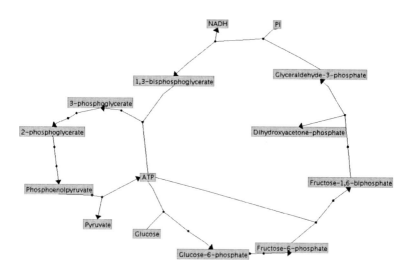

Figure 4.22.: Visualisation of glycolysis (2) done by the new layout algorithm. In contrast to figure 4.21 all compounds are treated as main compounds. The result are two cycles joined at ATP. This example shows the importance of the choice of side compounds and demonstrates their influence on the topology of a pathway.

4.4. Layout comparison

Since I could use the algorithms of Becker et al. and the Karp et al. with the same set of pathway, I created a scoring function to compare the results according to the handmade drawings from textbooks. In the first column of the tables the results of the existing layout algorithm are shown and in the second the results of the new algorithm. Furthermore, in the second and third rows are the results of the used scoring function. This function consists of two parts. The first part counts edge crossings (one crossing = 1 point). The second part gives points in the following way according to the detection of cycles:

- no cycle of the static drawing was found or a cycle was found although the pathway is hierarchical = 3 points

- a cycle was partly found = 2 points

- a cycle was completely found = 0 points

The algorithm with least points shows the pathway as most similar to the given handmade example. The best score is zero which means that the dynamic layout is similar or equal to the static graph representation. All handmade examples (except the PO reaction) can be found in the biochemistry book [136]. The order of examples corresponds to the increasing complexity (e.g. increasing node connectivity). The bar charts in the third column beneath the handmade drawing show the connectivity of the compounds (number of reactions per compound).

4.4.1. Becker et al. examples

These section contains the comparison with the algorithm of Becker et al.. The score for each example and algorithm is summarised in table 4.5. As you can see in this table the new layout algorithm has a score of only four points which is 15 times lower than the score of the algorithm of Becker et al.. This scoring proves that the new algorithm improves the layout results by reducing edge crossings (e.g. table 4.13) and edge lengths in the resulting graphical representation. It also considers existing biological conventions and creates a drawing that are very similar to the handmade examples from textbooks biochemists are familiar with.

Example	Becker et al.	new algorithm
Table 4.6	0	0
	3	0
Table 4.7	0	0
	0	0
Table 4.8	0	0
	0	0
Table 4.9	3	1
	2	0
Table 4.10	2	0
	6	0
Table 4.11	0	0
	3	0
Table 4.12	2	1
	3	0
Table 4.13	30	2
	6	0
Sum	61	4

Table 4.5.: Summary of the comparison between the algorithm of Becker et al. and the new layout algorithm.

4. Visualising biochemical pathways

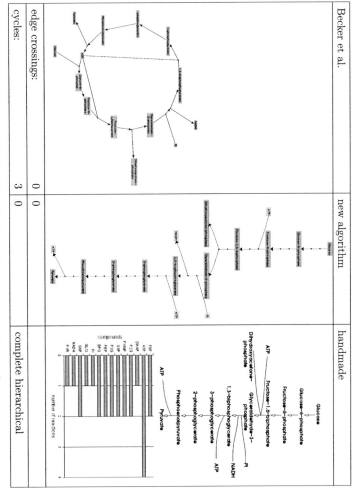

Table 4.6.: Glycolysis (Becker et al. and new layout algorithm).

70

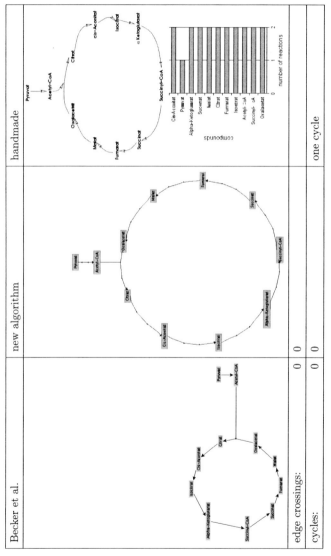

Table 4.7.: Citrate cycle (Becker et al. and new layout algorithm).

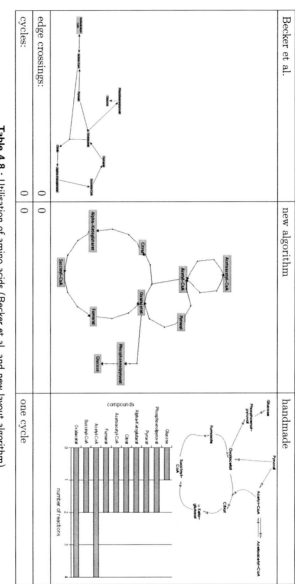

Table 4.8.: Utilisation of amino acids (Becker et al. and new layout algorithm).

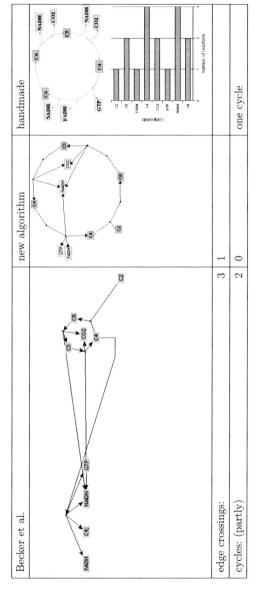

Table 4.9.: Overview of the citrate cycle (Becker et al. and new layout algorithm).

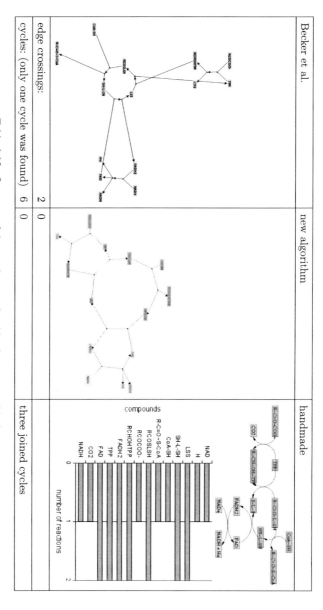

Table 4.10.: Summary of the reactions catalysed by the pyruvate dehydrogenase complex (Becker et al. and new layout algorithm).

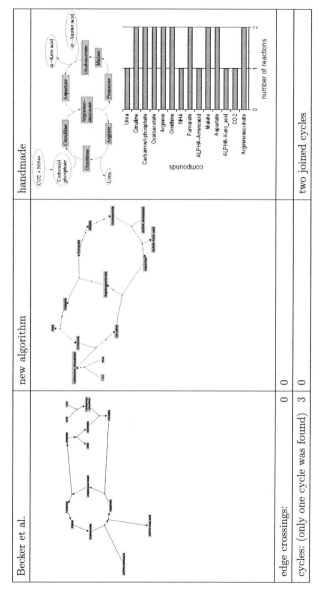

Table 4.11.: Urea cycle and parts of the citrate cycle (Becker et al. and new layout algorithm).

Becker et al.		new algorithm	handmade
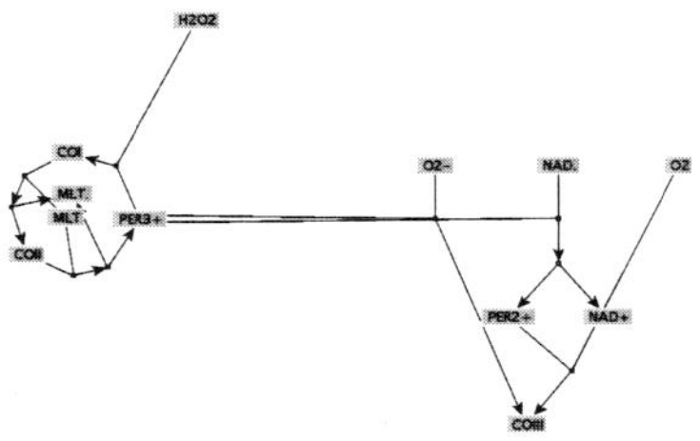		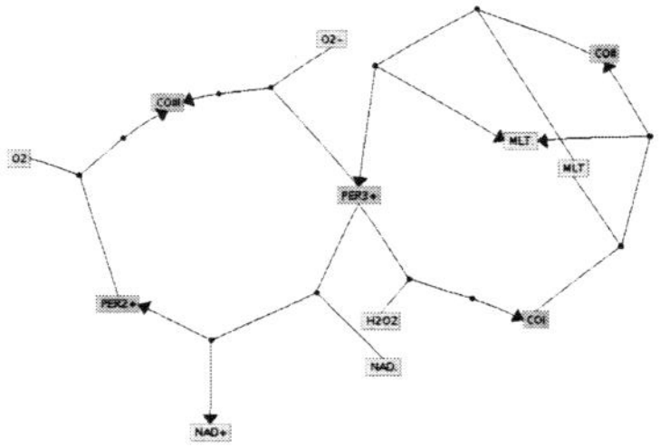	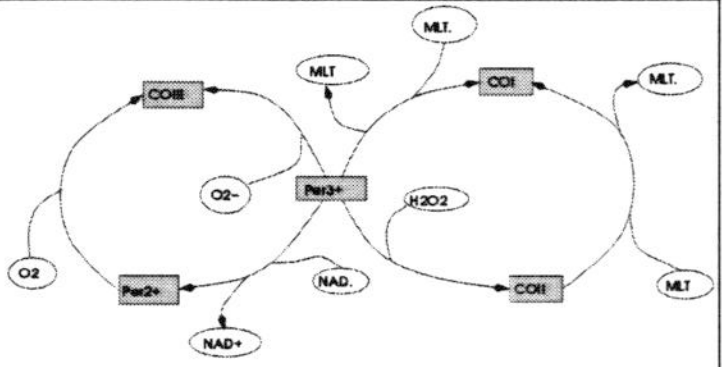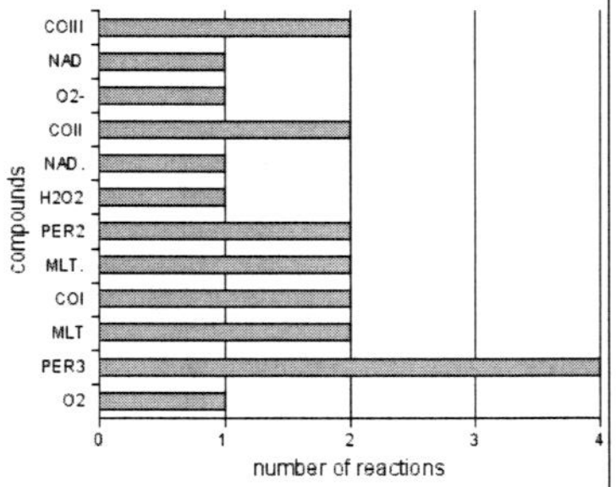
edge crossings:	2	1	
cycles:	3	0	two cycles

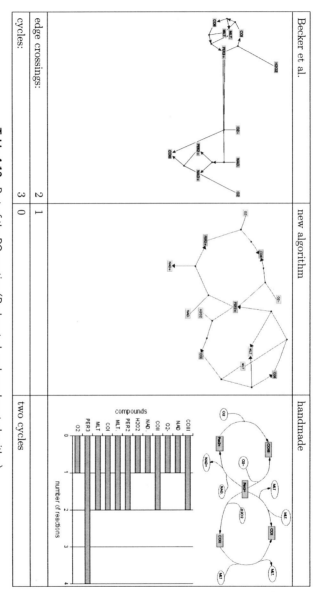

Table 4.12.: Part of the PO reaction (Becker et al. and new layout algorithm).

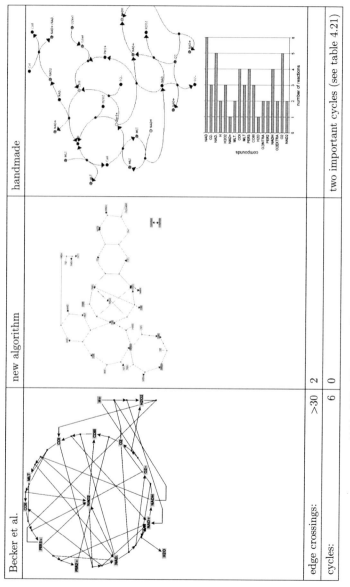

Table 4.13.: PO reaction (Becker et al. and new layout algorithm).

4. Visualising biochemical pathways

4.4.2. BioCyc examples

The score for each example done by the Karp et al. and the new algorithm is summarised in table 4.14. As you can see in this table the new layout algorithm has a score of only four points which is eight times lower than the score of the algorithm of Karp et al.. This scoring proves again that the new algorithm improves the layout results of complex biochemical pathways.

Example	Karp et al.	new algorithm
Table 4.15	0	0
	0	0
Table 4.16	0	0
	0	0
Table 4.17	0	0
	0	0
Table 4.18	1	1
	0	0
Table 4.19	3	0
	6	0
Table 4.20	0	0
	3	0
Table 4.21	2	1
	3	0
Table 4.22	8	2
	6	0
Sum	33	4

Table 4.14.: Summary of the comparison between the algorithm of Karp et al. and the new layout algorithm.

78

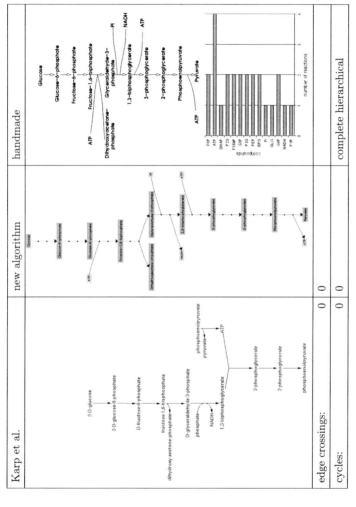

Table 4.15.: Glycolysis (Karp et al. and new layout algorithm).

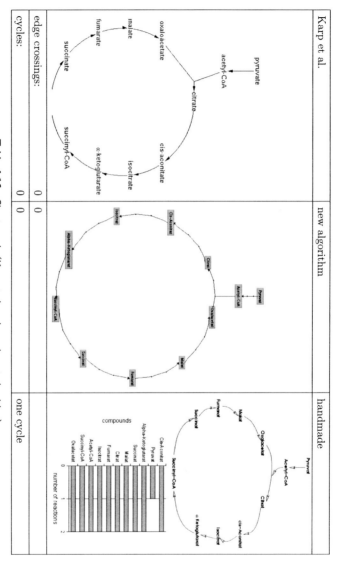

Table 4.16.: Citrate cycle (Karp et al. and new layout algorithm).

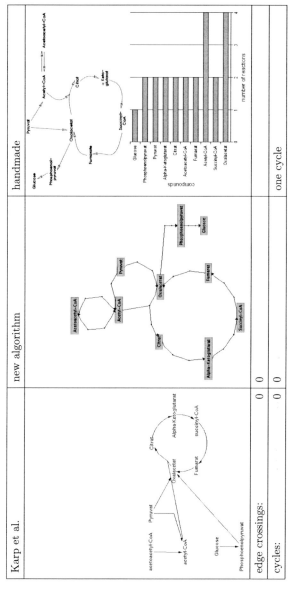

Table 4.17.:: Utilisation of amino acids (Karp et al. and new layout algorithm).

4. Visualising biochemical pathways

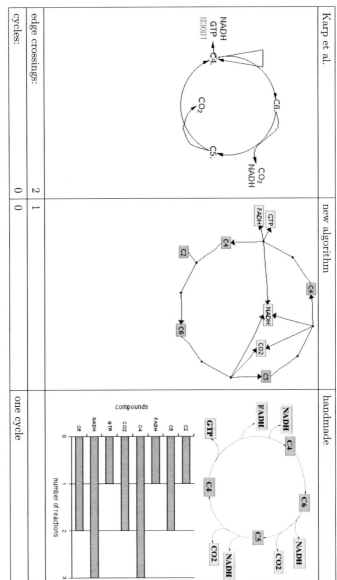

Table 4.18.: Overview of the citrate cycle (Karp et al. and new layout algorithm).

	Karp et al.	new algorithm	handmade
edge crossings:	2	1	0
cycles:	2	0	one cycle

82

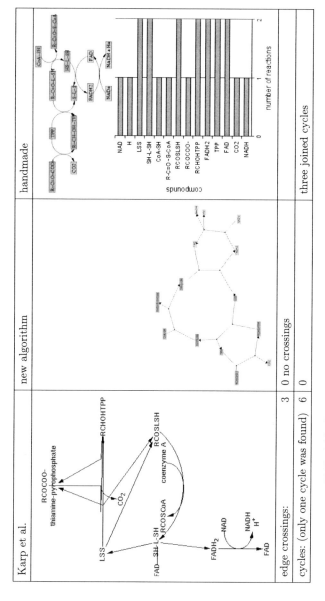

Table 4.19.: Summary of the reactions catalysed by the pyruvate dehydrogenase complex (Karp et al. and new layout algorithm).

4. *Visualising biochemical pathways*

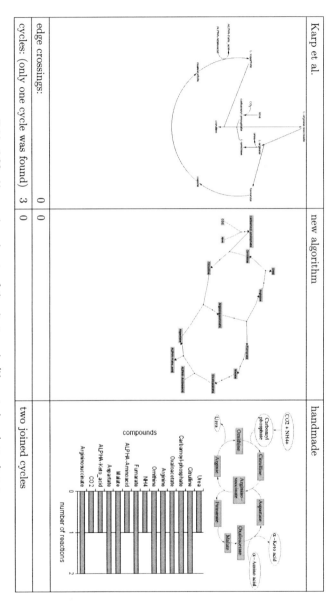

Table 4.20.: Urea cycle and parts of the citrate cycle (Karp et al. and new layout algorithm).

84

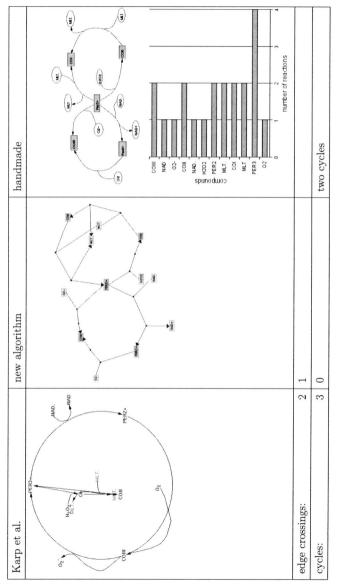

Table 4.21.: Part of the PO reaction (Karp et al. and new layout algorithm).

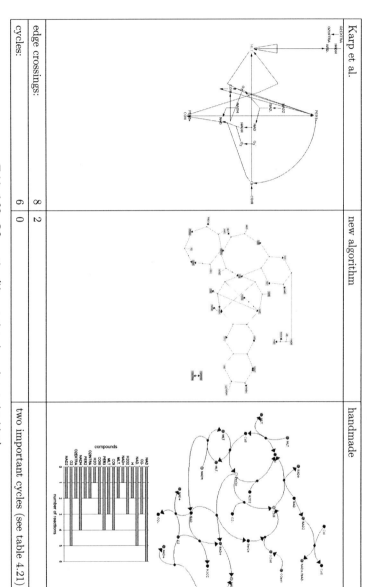

	Karp et al.	new algorithm	handmade
edge crossings:	8	2	
cycles:	6	0	two important cycles (see table 4.21)

Table 4.22.: PO reaction (Karp et al. and new layout algorithm).

4.5. Implementation

4.5.1. The existing algorithm of Becker et al.

As a first step towards the implementation of the new layout algorithm proposed, a revision of the class diagram of the algorithm of Becker et al. and its implementation is carried out. This was done due to the fact that the new algorithm is based on the one proposed by Becker et al.

Becker et al. use the YFiles library [135] for the graph layout. This library offers classes to create and handle graphs and their layouts. Additionally, this library is written in Java which means platform independence. At the beginning of this project there was no other Java library that offered data structures for representing graphs, nodes or edges as well as hierarchical and circular layout algorithms, thus I decided also to use this library.

The implementation of the algorithm of Becker et al. defines, apart from three general utility classes (Node/EdgeCursorImplementation, Geometry), three main classes, namely PathwayLayouter, Subgraph, and ForceDirectedLayout (see the class diagram in figure 4.23).

In the class ForceDirectedLayout Becker et al. implemented a force-directed algorithm that takes the node width and height and a preferred edge length into account. The class Subgraph contains attributes (a reference of the graph to which this subgraph belongs and a list of nodes that are in this subgraph) and access methods (nodes(), edges(), incomingEdges(), outgoingEdges() or moveTo()). Consider as an example, the moveTo() method moves a subgraph to a new position by adding the given parameter dx and dy to the current x- and y-coordinates of this subgraph. Furthermore, it contains methods to find connected or strongly connected components, as well as a DFS implementation to find the longest cycle in this subgraph.

The class PathwayLayouter coordinates the search for circular and hierarchical subgraphs in a given graph. Firstly, it looks for all cycles in the graph, and then saves the longest of them. This step is repeated with all remaining nodes that are not part of a cycle until no more cycles can be found. Secondly, all nodes that are not part of a cycle are joined in one hierarchical subgraph, whether they are connected or not. Each subgraph (cycle or hierarchical subgraph) are objects of

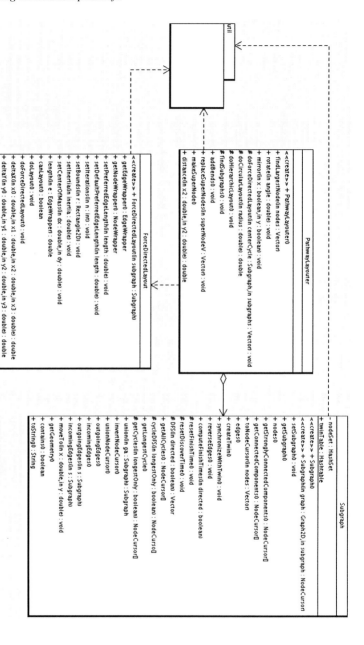

Figure 4.23.: Class diagram of the existing layout algorithm of Becker et al. (see the classes of the package util in figure 4.24).

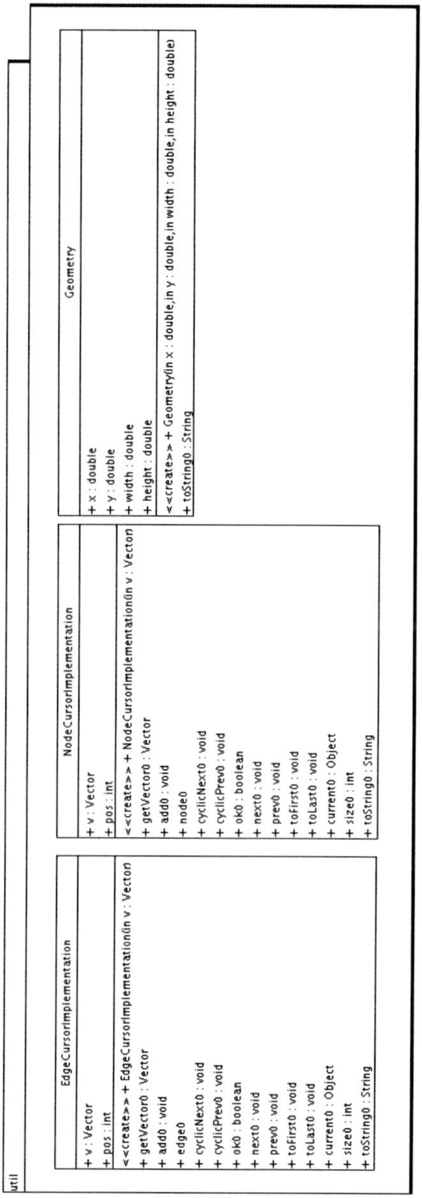

util

EdgeCursorImplementation

+ v : Vector
+ pos : int

<<create>> + EdgeCursorImplementation(in v : Vector)
+ getVector() : Vector
+ add() : void
+ edge()
+ cyclicNext() : void
+ cyclicPrev() : void
+ ok() : boolean
+ next() : void
+ prev() : void
+ toFirst() : void
+ toLast() : void
+ current() : Object
+ size() : int
+ toString() : String

NodeCursorImplementation

+ v : Vector
+ pos : int

<<create>> + NodeCursorImplementation(in v : Vector)
+ getVector() : Vector
+ add() : void
+ node()
+ cyclicNext() : void
+ cyclicPrev() : void
+ ok() : boolean
+ next() : void
+ prev() : void
+ toFirst() : void
+ toLast() : void
+ current() : Object
+ size() : int
+ toString() : String

Geometry

+ x : double
+ y : double
+ width : double
+ height : double

<<create>> + Geometry(in x : double, in y : double, in width : double, in height : double)
+ toString() : String

Figure 4.24.: Class diagram of the util package of Becker et al. (see also figure 4.23).

the class Subgraph. The PathwayLayouter then layouts each subgraph according to
its topology calling the method doCircularLayout() that uses the class SingleCycle-
Layouter of the YFiles library or the method doHierarchicalLayout() that uses the
class HierarchicLayouter of the YFiles library. Finally, the subgraphs are reassembled to a whole graph with the aid of the ForceDirectedLayout class. Therefore,
the PathwayLayouter creates for each subgraph a node with the width and height
of this subgraph using the method makeSuperNode(). After the force-directed layout the subgraphs are moved to the position of the corresponding node (method
replaceSuperNode()).

4.5.2. The new layout algorithm

The architecture of PathWiz

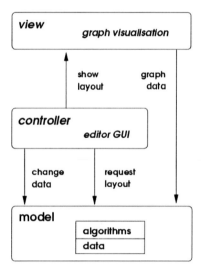

Figure 4.25.: The MVC architecture of the PathWiz module. The italic parts were
implemented by Dr. Ursula Rost.

As mentioned in section 3.6 ViPaSi is based on the model - view - controller
concept. This section shows how PathWiz module has been integrated into the

ViPaSi architecture.

The PathWiz module contains a view of a graph and user interactions in a graph editor developed by Dr. Ursula Rost (view and controller). Within the scope of this thesis I implemented the new layout algorithm which is part of the model (see figure 4.25). Therefore, I will introduce only the implementation of the new layout algorithm in more detail.

The user loads a file or creates a graph in the graph editor that changes the pathway and layout information in the model. If the user requests a layout, the controller will send a layout request to the model. If the layout is done, the controller will send an update request to the view that reads the new layout information from the model and displays it. The model part should be split into two sub-parts. The data part contains data structures to describe graphs with nodes and edges as well as subgraphs. The algorithms part consists of algorithms related to investigate or layout the graph.

Since the YFiles library and the class Subgraph contain both data structures and algorithms, I could not separate clearly the data and the algorithm part as mentioned above. Therefore, the model contains the YFiles library, the classes Subgaph, Node- and EdgeCursorImplementation in the data part and ForceDirectedLayout, PathwayLayouter and the classes for the new layout algorithm are in the algorithms part, see the realized architecture in figure 4.26.

The classes of the new algorithm

The mentioned Becker et al. classes ForceDirectedLayout, PathwayLayouter and Subgraph are reused in the implementation of the new algorithm in order to find connected components, to create circular or hierarchical layouts and to reassemble found subgraphs. However, new classes and methods were also required (see an overview in figure 4.27).

The new algorithm classes ComplexPathwayLayouter and ComplexSubgraph are inherited from the Becker et al. classes PathwayLayouter and Subgraph to reuse existing methods. Furthermore, new classes and methods are needed that

- provide the BFS for searching the smallest cycle,

- split and join nodes,

91

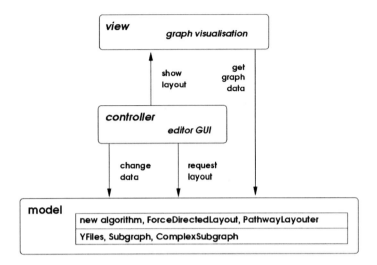

Figure 4.26.: The realized architecture of the PathWiz module.

- calculate edge crossings,

- handle side and main compounds, or

- handle layout options.

Figure 4.27 shows the class diagram of the new algorithm distinguishing between the classes that were taken from the Becker et al. implementation and the new designed and implemented classes. The class diagram in figure 4.28 shows the important methods and attributes of the new created classes.

All classes in figure 4.28 are in the model. The classes LayoutOptions, ListOf-SideCompounds, and ComplexSubgraph belong to the data structures and the other classes to the algorithms part.

The ComplexSubgraph contains a BFS implementation that searches for the smallest cycle or for the shortest path between two nodes. The existing DFS method finds the longest cycle since it first looks in the depth of the graph instead of the breadth. Furthermore, I added methods that check if the found cycle meet all requirements and methods that add or remove a node when nodes are split or

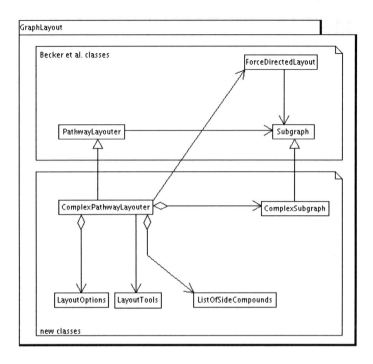

Figure 4.27.: Class diagram of the implemented new layout algorithm separated in existing (algorithm of Becker et al.) and newly created classes.

joined.

The class ComplexPathwayLayouter is the main class of the new algorithm and coordinates the layout process. This class consists of the most important methods of the new layout algorithm. I designed only one class because if I would have implemented several classes, there would be a huge amount of data structures that would have to be shared and synchronised between these classes.

The class ComplexPathwayLayouter also contains the method findSubgraphs() described in section 4.3.1. This method firstly looks for circular subgraphs and secondly splits the remaining nodes that are not part of a cycle into connected components that represent hierarchical subgraphs. Each found subgraph is an object of the class ComplexSubgraph and is saved in the corresponding list (found-

4. Visualising biochemical pathways

Cycles, foundHiers). Circular subgraphs are layouted in the existing method doCircularLayout() by using the SingleCycleLayouter class in YFiles.

The layout of a hierarchical subgraph needs some preparations beforehand. Firstly, all split nodes in the hierarchical subgraph are joined. Secondly, the nodes are separated into main and side compounds. Thirdly, the main compounds and all dummy nodes representing reactions are layouted by the HierarchicLayouter class using the Becker et al. implementation. Finally, each side compound is split into as many nodes as it has edges. The resulting nodes with degree one are placed one layer above or under the corresponding end node according to the direction of the edge (top to down).

If all subgraphs are found, the class ForceDirectedLayout of the algorithm of Becker et al. reassembled these subgraphs to a whole graph, but the centre cycle is not the circular subgraph with the largest number of nodes as in the algorithm of Becker et al.. In the new algorithm the centre cycle is the subgraph with the largest number of edges to other subgraphs because the ForceDirectedLayout places the other subgraphs around this centre cycle and edge crossings between subgraphs are reduced when the highest connected subgraph is in the centre. Of course, in many cases the highest connected subgraph is also the circular subgraph with the largest number of nodes.

After placing the subgraphs the method checkEdgeCrossings() in the class ComplexPathwayLayouter reduces edge crossings by using the method countIntersections(). The class Intersection calculates the intersection point of two edges. The two edges are represented by objects of the class LineSegment. The method countIntersections() is part of the class LayoutTools. This class is an utility class that is not a real class but a collection of additional methods [137] for the layout algorithm that can be used by different classes. The class diagram in figure 4.28 only shows the most important methods of this class. The method joinTwoCycles() moves two circular subgraphs next to each other at the position of a split node and joins this node in the first subgraph and its splitting partner in the second subgraph. The method switchYCoordinates() switches the top to down hierarchy of a hierarchical layout into a down to top hierarchy according to edge directions. This method will be used if a hierarchical subgraph has two crossed edges to another subgraph.

The class ListOfSideCompounds contains a list of compound names that are considered as side compounds during the layout process. Since the list is changeable by the user, I implemented methods to add and remove compound names from the list. This class is designed as a Singleton design pattern which guarantees that there exists only one instance of this class and therefore all classes use the same list.

The class LayoutOptions saves the user options for the layout process:

- the number of compounds in the smallest cycle (default = three),

- the number of allowed edge crossings per edge (default = two),

- whether the algorithm shall differentiate between main and side compounds (default = true),

- whether each side should be represented by one node per occurrence in a reaction (default = false), and

- whether the number of dummy nodes should be minimised (default = false).

For example, as default each reaction is represented by two dummy nodes, but if the option minimiseDummyNodes is true, the number of dummy nodes per reaction depends on the number of substrates and products in this reaction (see table 4.23). In the first case, if the reaction has one substrate and one product, there will be no dummy node. In the second case, if the reaction contains one substrate and more than one products or more than one substrates and one only product, it has one dummy node. In all other cases each reaction will have two dummy nodes. This minimisation reduces the number of nodes in the graph and this reduces the time needed by the algorithm to layout the given graph. The coordinates of the missing nodes are calculated after the layout process.

The connection of the layout classes to the graph editor

The figure 4.29 displays the classes that connect the new layout algorithm with the graph editor in the PathWiz module. The editor contains objects of the classes OptionsDialog and ListDialog that are part of the controller like the editor.

The user sets the layout options through the OptionsDialog box that creates an object of the class LayoutOptions. Since the number of compounds in the smallest cycle and the number of edge crossings are set by the user. They are represented

Substrates	Products	Dummy nodes	
1	1	0	Fumarate ————▶ Malate
1	> 1	1	Argininosuccinate ⟨ Fumarate / Arginine
> 1	1	1	Citrulline ＼ / Aspartate ▶ Argininosuccinate
> 1	> 1	2	α–Keto acid ＼ / Oxaloacetate ▶ Aspartate / α–Amino acid

Table 4.23.: Creating of dummy nodes. If the option minimiseDummyNodes is set to true, the number of dummy nodes created per reaction in the layout graph depends on the number of compounds in the current reaction as displayed in this table.

by editable java.swing.JTextField objects. The other options have as value only true or false. Therefore, they are designed as radio buttons (java.awt.ButtonGroup) that allow only one choice.

The ListDialog box enables the user to edit the list of side compound names (ListOfSideCompounds), e.g. add or delete compound names. These actions are implemented as buttons, e.g. the add button opens a small dialog box that shows a text field to enter a compound name.

When the user requests a layout, the graph editor starts the layout process by calling the doLayout() method of the ComplexLayouter class in the model part.

The class ComplexLayouter uses the vector of GraphicalReaction objects, the current LayoutOptions and ListOfSideCompounds object to prepare the layout process. The class ComplexLayouter connects the YFiles graph structure with the reaction information in the vector of GraphicalReaction objects. These reactions are converted into nodes and edges in a Graph2D object of the YFiles library. Firstly it creates for each compound a Node object with the height and the width of the corresponding compound. Secondly, it produces dummy nodes representing reactions and thirdly it produces the Edge objects connecting the compounds with the dummy nodes. Finally, the Graph2D object is send to the class ComplexPathwayLayouter that creates the layout for this graph. After layouting the ComplexLayouter saves the new coordinates into the vector of GraphicalReaction objects. If a node is split, the ComplexLayouter will create a new GraphicalNode that replaces its splitting partner in the according reactions. The editor informs the view that will then read the modified vector of GraphicalReaction objects and visualise the new layout on the screen.

With the main method in the class ComplexLayouter the user can use the new algorithm as command line tool that reads the pathway information from an SBML file and adds the layout results into this file.

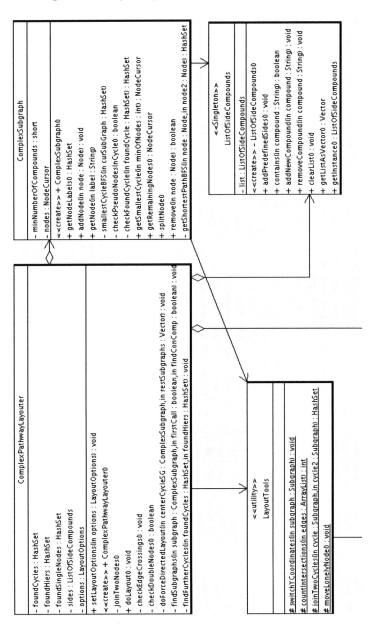

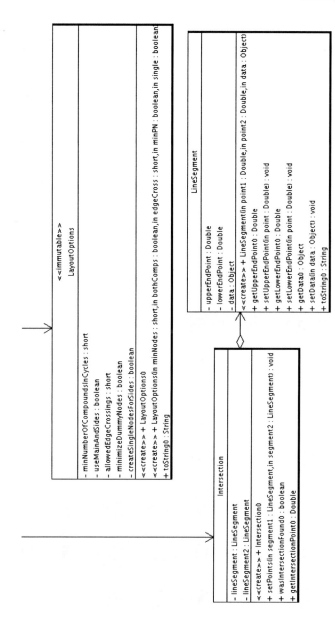

Figure 4.28.: Detailed class diagram of the new classes implementing the new layout algorithm.

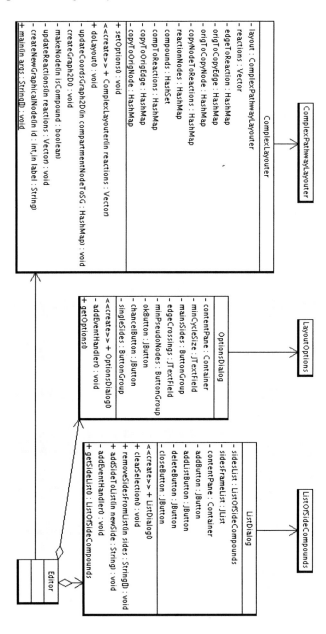

Figure 4.29.: Class diagram displays the classes that connect the layout algorithm to the graph editor application.

4.5.3. Complexity

Finally, some words about the complexity of the new layout algorithm. The bottleneck is the cycle search using the breadth first search (BFS) which is the standard method for finding cycles and guarantees that the shortest path between two nodes is found [68].

In the best case (figure 4.30), the BFS starts with a node that is part of a cycle with the required number of

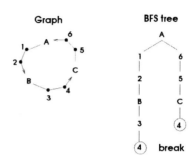

Figure 4.30.: An example of the best case.

compounds and there are no other side branches in the tree until the last node of the cycle is reached. Then, the BFS finds this cycle immediately in constant time $(O(2(M + DN))$, M = number of nodes in the smallest cycle, DN = number of dummy nodes representing participating reactions, 2 = the search visits the same number of edges as nodes).

If the graph does not contain any cycle at all, or if the existing cycles are smaller or larger than the defined smallest cycle, a complete breadth first tree will be calculated. According to this, there are some cases in which the complexity of the search will be exponential $(O(2^{N+E}))$ since in every third level the number of nodes will be doubled. Figure 4.31 shows an example of the worst case.

The calculation of the BFS tree will be stopped if the maximum depth is reached. The maximum depth corresponds to the number of nodes in the graph because if all nodes of the current graph are part of one big cycle, the smallest cycle will contain all nodes and could only be found if the depth N is reached. Since this depth limit does not guarantee that every cycle is found, such a tree is created for every compound with at least two edges.

However, the worst case occurs very rarely in biochemical pathways. This is due to several reasons. Firstly, in most pathways a cycle with the required number of nodes is found which has a much lower complexity leading to faster results. Secondly, if any cycle exists in a branch, this branch will not be continued which reduce the complexity, too. Thirdly, the expansion of a tree will be stopped, if a valid cycle is found. Finally, in cases where the BFS tree is completed, the specific

4. *Visualising biochemical pathways*

structure of the pathway will not lead to exponential running time in most cases.

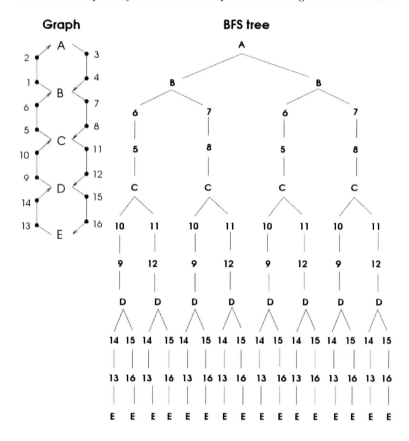

Figure 4.31.: An example of the worst case of the implemented BFS.

5. Visualisation of biochemical simulation results

> *"Scientific visualisation is exploring data and information graphically, gaining understanding and insights into the data"*
>
> *(R. A. Earnshaw, 1973)*

Modelling and simulation which are used more and more frequently to support experimental investigations are leading to large amounts of times series data. Therefore, it becomes increasingly important to develop visualisation tools which facilitate the analysis of the respective data. This chapter will introduce the ViPaSi module SimWiz3D which visualises time series data in a concise way using a three-dimensional representation and provides a wide range of user interactions.

5.1. Scientific visualisation

Before motivating the new software module SimWiz3D some words about scientific visualisation. In addition to the definition of scientific visualisation by Earnshaw [55] at the beginning of the chapter, I am citing Robinson et al. [138] who formulated the task of scientific visualisation:

> *Scientific visualisation is more than just "pretty graphics". If visualisation is to be useful it must shed new light on data and lead to previously unknown, but valid and useful conclusions, provide means to manipulate and browse data more easily, or provide intuitive user interfaces that increase productivity.*

Since biochemical networks under studies are increasing, more and more species are part of a model and therefore, simulation results have a high dimensionality, and suitable visualisation tools are necessary that support the understanding of these networks. Firstly, I will discuss visualisation techniques for time-dependent data in general and then I will introduce visualisation tools specifically for biochemical simulation results.

5.1.1. Visualisation of time-dependent data in general

By using visualisation tools, the user wants to answer questions about his data. In the case of time-dependent data MacEachren [139] set up the following list:

- *Does a data element exist at a specific time?*

- *When does a data element exist on time?*

- *How fast is a data element changing or how much difference is there from data element to data element over time?*

- *In what order do data elements appear?*

- *Do data elements exist together?*

Therefore, a suitable visualisation of time-dependent data should answer these questions by detecting temporal patterns and temporal behaviour of data elements [140]. There are a lot of visualisation techniques for time-dependent, multivariate data which are listed below (reviews: [140] and [141]).

Conventional techniques: Time series are mostly visualised as line graphs (x-coordinate = time, y-coordinate = time series values). A line graph is a plot in which the data points are connected by lines [142].

A scatterplot shows a line graph of two time series (x = time series values of variable 1, y = time series values of variable 2) and allows the visual detection of linear correlations. Scatterplot matrices are used to compare scatterplots of each time series pair in the data, but e.g. for 20 time series the matrix contains 190 scatterplots which cannot be observed well by the user.

Calendar view [143]: This technique was developed to visualise clusters of daily data. It contains two views. The first view shows a calendar plot (days and months) in which the days are colour-coded according to a cluster method. In the second view the data of a chosen cluster, day, or month is visualised as line graph. The time scale in simulation data is not restricted to days and month and can also be in ms.

Lexis Pencils [144]: Different time series are mapped to different faces of a pencil. Several pencils can be presented together by locating them in 3D to the spatial context of the data. Since the faces are limited, it is not suitable for a large number of dimensions. Furthermore, according to the orientation of a pencil, several faces are hidden and cannot be compared to other faces.

MultiComb [145]: Each time series is represented by one axis. All axes are arranged in a cycle like a star outward from a centre. The centre can be used for visualising further information. Since the data is circular arranged, it is hard to compare values in data sets that are placed on opposite sides of the circle.

Spiral Clock [146]: Cyclic data is visualised as a clock. By moving the mouse over the clock for- and backwards the data changes automatically to the chosen time range.

Spiral Graph [147, 148]: The data is mapped to a spiral ring using colour, texture and thickness. This technique is flexible as well as suitable to detect periodic structures in time series data.

TimeWheel [145]: The time axis is situated in the centre of a wheel and the other axes representing the time series are arranged on a circle around it. Since a line connects a time step with a value in a time series axis, there are a lot of line crossings which make it difficult to analyse the data.

VizTree [149]: The time series are discretised to symbols. These symbols are encoded into a modified suffix tree. Each branch of this tree represents a subsequence pattern of the data. The thickness of this branch corresponds to the frequency how often this pattern was found in the observed data.

Wormplots [150]:   Wormplots are based on scatterplots. For each time step scatterplots are created. In the first scatterplot point clouds are identified. These clouds are represented by circles. The defined clouds are identified in all other scatterplots of the remaining time step and these circles are connected like a worm. Therefore, the temporal behaviour of the clouds can be analysed.

These techniques are able to visualise time-dependent data, but they are mostly restricted to a special kind of data (e.g. cyclic data - SpiralClock, SpiralGraph, TimeWheel) or are limited in the number of time series that can be visualised in a suitable way (e.g. line graphs, Lexis Pencils). Some tools search for special patterns in the data (e.g. SiralGraph, Wormplots), but that would assume that researchers know which patterns they are looking for which is mostly not the case for simulation data of an unknown system.

Furthermore, they would not answer all questions about biochemical simulation results. Biochemists are interested in a continuous visualisation of the concentration changes of the species in a pathway over time and the correlation between them. Therefore, they are asking the following questions:

- How do reactions or complete pathways work?

- How do they work together?

- Are there correlations and which are these?

Since some correlations depend on the connectivity of reactions or species in the pathway which can be seen in the graph representation of this pathway, a visualisation is necessary that is able to show both time series data and the graph representation of the pathway. None of the above methods is able to visualise the graph representation.

There are already some tools that are able to show the time series and the graph representation simultaneously. These tools will be briefly described in the next section.

5.1.2. Visualisation of biochemical simulation results

In most simulation tools (e.g. table 2.3) these biochemical time series are represented as line graphs. In the case of simulation data the x-axis represents the time

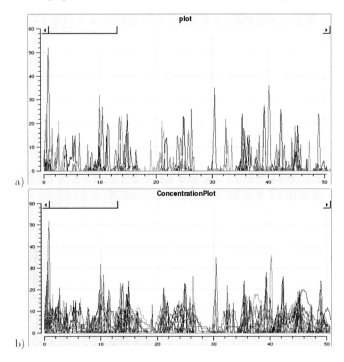

Figure 5.1.: Concentration-time-plots, a) contains 10 species and b) 30 species.

and the y-axis the concentration or particle number of each species in the pathway but these concentration-time plots are only suitable for a small number of reactants. On the one hand the more curves that are present in the same diagram the harder it is to follow each curve, even if different colours are used (figure 5.1). Figure 5.1 shows two concentration-time plots. The first plot contains 10 species and the second 30. In the first plot it is already hard to differentiate between the 10 species, but in the second plot you cannot see anything. On the other hand the representation is inappropriate when the range of concentration values varies largely, e.g. curves with a smaller range could look like a flat line, although they exhibit oscillations.

For these reasons it is necessary to develop visualisation techniques to facilitate the handling and visual analysis of multiple time series. This will improve the understanding of the studied processes.

As mentioned already in the previous section the goal is a visualisation which takes the time series and the graph representation into account which is achieved by all following tools.

Talis

Talis contains a step-by-step visualisation of simulation results. Each image represents one time step. The simulation results are mapped to the graph representation of the pathway (figure 5.2). Each species in the graph representation has a bar. The length of this bar corresponds to the maximum value of this species. The bar inside this bar indicates the current value at the shown time step. By animating the results the visualisation is changed automatically time step by time step.

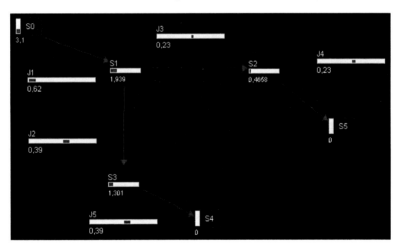

Figure 5.2.: Screen shot of Talis. The length of the bar of each species corresponds to the maximum value and the short bar inside this bar corresponds to the current value of the shown time step.

SimWiz

SimWiz as part of ViPaSi was already introduced in section 3.4. It contains a shape-coding and a colour-coding step-by-step visualisation of simulation time series data (figure 5.3). An animation feature is also integrated.

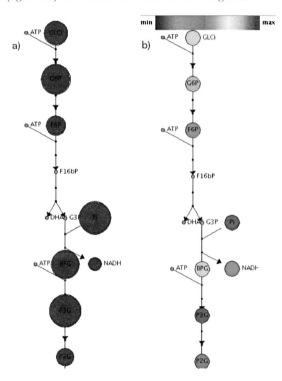

Figure 5.3.: Screen shot of SimWiz. It shows one time step of a part of the glycolysis pathway. a) shape-coding (the diameter of the circles corresponds to the concentration values). b) colour-coding (the colour corresponds to the concentration value).

5. Visualisation of biochemical simulation results

Borisjuk et al.

This visualisation tool offers two techniques. The first technique maps the concentration values of one time step of different simulations on the graph structure by using bars (figure 5.4 a)). This technique has the advantage that different simulation settings can be compared. The second technique maps small concentration-time plots of two simulation settings on the graph representation (figure 5.4 b)).

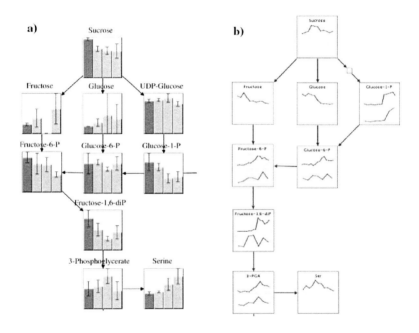

Figure 5.4.: Visualisation of simulation results in the Borisjuk et al. tool.

Unfortunately the concentration-time plots are rather small and it is hard to compare species that are far away in the graph and to compare the ranges of the concentration values. Furthermore, you cannot recognise well how long a maximum or minimum lasts in the series or which species are at their maximum or minimum at the same time. It is a nice overview visualisation but is not suitable for a good exploration and interpretation of the data.

To summarise

All three tools contain a step-by-step visualisation which shows one time step of the simulation in one image. The continuity of the data is completely lost. Because of the animation feature in SimWiz and Talis the user could get an impression of this continuity for rather short time periods, but e.g. correlations about long intervals are hard to be recognised. Therefore, visualisations which show complete time series are necessary. The second visualisation technique in Borisjuk et al. shows complete time series but they are very small and therefore not suitable for interpreting the results.

Dwyer et al.

At the moment there exists one approach by Dwyer et al. [151] for a more sophisticated visualisation of complete time series data. This tool visualises the reaction network in the x,y-plane and uses the third dimension z to represent experimental results. It is possible to switch between a disc or a box (similar to a histogram) representation. For each time step it creates a single disk (figure 5.5) or box (figure 5.6) and displays them in a consecutive manner.

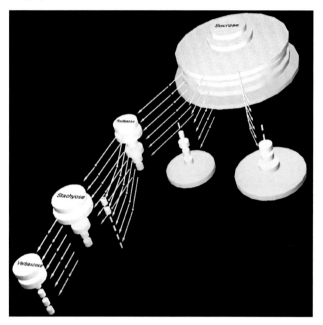

Figure 5.5.: Visualisation of simulation results by using discs in the Dwyer et al. tool.

Unfortunately, in this representation the continuity of time series is also lost a little because each time has its own disc or box representation and there is no smooth transition between the steps. Furthermore, it is hard to compare single time steps or complete time series, especially in the box representation if the box length is vertical to the users' point of view (e.g. galactose in figure 5.6). Arrows displaying the flow in the network for each step are also disconcerting. Finally, the user has to write a GML file (Graph Modelling Language [152]) including the network structure and the corresponding time series that costs time.

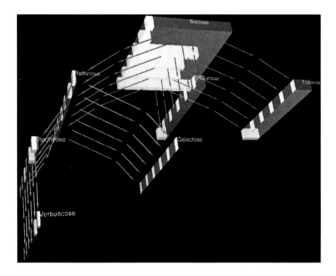

Figure 5.6.: Visualisation of simulation results by using boxes in the Dwyer et al. tool.

Because of these disadvantages and the lack of user interactions that Johnson [153] described as key feature of a visualisation tool, I developed a new visualisation tool: SimWiz3D. Next, I will specify the demands of the tool and describe the tool itself in detail.

5.2. Software demands

Shneiderman defined visual design guidelines that are called together the Visual Information Seeking Mantra: *Overview first, zoom and filter, then details-on-demand* [154]. It consists of the following seven tasks:

5. *Visualisation of biochemical simulation results*

- *Overview*

- *Zoom*

- *Filter*

- *Details-on-demand*

- *Relationships*

- *History*

- *Extraction*

Carr [155] supplemented further demands on a visualisation tool:

1. *User tasks must be supported.*

2. *The graphic method should depend on the data.*

3. *Navigation and zooming do not replace filtering.*

4. *Multiple views should be coordinated.*

Since visualisation is a *internal mental process* [142], the new visualisation tool should also take the human perceptual capabilities into account. The working of the eyes is well studied and therefore, we know that colour, texture and relative positions can be realized very fast [156]. When visualising 3D structure shadows, textures and movements are most important for the perception [156]. Furthermore, the human response time to varying visual stimuli should be considered [157].

After introducing SimWiz3D in detail I will analyse it according to these tasks. Firstly, SimWiz3D needs to organise, transfer, manipulate and store the data. Secondly, the visualisation has to support the understanding of the data and to emphasise relationships. Thus, a lot of user interaction should be integrated, so that the user can adapt the tool to his demands. According to the visualisation pipeline in figure 5.7 the user should be able to interact with the filtering, mapping, and rendering process.

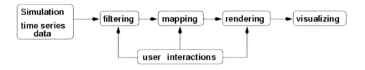

Figure 5.7.: Visualisation pipeline.

5.3. SimWiz3D

SimWiz3D combines the step-by-step visualisation and the animation feature of the SimWiz module with a new 3D-View of the time series data and further user interaction and data exploration techniques.

5.3.1. 3D-View

Similar to Dwyer et al. the 3D-View visualises the read layout information of the graph structure in a 2D plane (x,y) and uses the third axis (z) as time (z=0 first time step). Therefore, it is not a homogeneous three dimensional visualisation, but a crossproduct of two distinct subspaces. The first space contains the network structure and the second the time series data over time. By using the third dimension this view can display a lot of time series simultaneously in a suitable way.

The time series are realized as tubes which is a technique that is used e.g. for fund data [158] or dynamic flow visualisations [142]. Nevertheless, tubes in dynamic flow visualisations are mostly a surface that is formed by several streamlines passing through a closed contour [159]. Such streamlines are poly-lines that indicate the direction of flows, e.g. flux in electromagnetic [142]. Sometimes these streamlines are also rendered as tubes. The thickness of these tubes corresponds to the velocity (the thinner, the faster). In contrast to this flow visualisation the tubes in SimWiz3D do not visualise any movements of species and the diameters of the tubes do not correspond to the velocities of the reactions, but to concentration or particle values at a certain time step.

In SimWiz3D each tube is composed of several conical frustums. The top radius of each conical frustum belongs to the time step n and the bottom radius to time step $n + 1$. The following conical frustum starts with time step $n + 1$ and ends with $n + 2$ and so on which creates a continuous appearance (see figure 5.8). Due to performance reasons, time steps with the same concentration values are joined to one conical frustum.

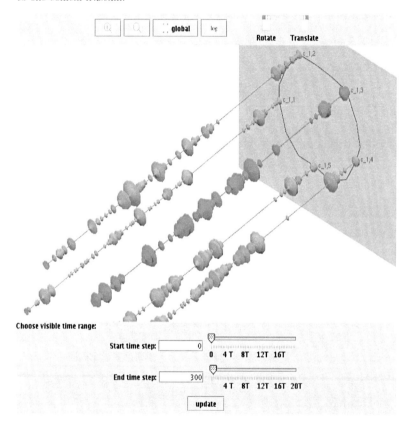

Figure 5.8.: 3D-View of biochemical simulation results and the cutting plane at a certain time step in SimWiz3D.

Similar to SimWiz and Dwyer et al. [151] the user can choose between two scaling types. The global scaling type takes the maximum concentration of all reactants as corresponding value for the maximum radius and the local scaling type the maximum concentration of each reactant. The maximum radius depends on the smallest distance between two nodes in the current network and corresponds to a quarter of this distance which prevents overlapping of tubes in the view. The minimum radius is by default one that makes a thin line and match the concentration or particle values equal to zero.

Furthermore, the user has the option to choose between two calculation methods. On the one hand the radius corresponds to the square-root of the concentration which means it is proportional to the area of the circle. On the other hand a logarithmic representation can be used which can be helpful if very different value ranges exist. These options can be chosen in the button line on the top of the 3D-View (see figure 5.8). The zooming buttons together with rotation and translation facilities allow the user to explore the 3D-View in an easy way. By double clicking on a tube this reactant and its reactions are highlighted in the 2D-View and a new small window will show the following additional reactant information (see figure 5.9):

- name of this reactant,

- statistical information:

  - start concentration value,

  - minimum concentration value,

  - maximum concentration value,

  - mean concentration value,

  - variance,

- button to switch the current colour of this tube,

- button to switch the transparency value of this tube.

This window is also available about the frame menu **Tool** by choosing a name in the submenu **Show statistics**.

With the start and end time slider found below the 3D-View the user can reduce the number of time steps displayed, to focus on interesting time ranges or to reduce data which would also increase the update response time when long time series are

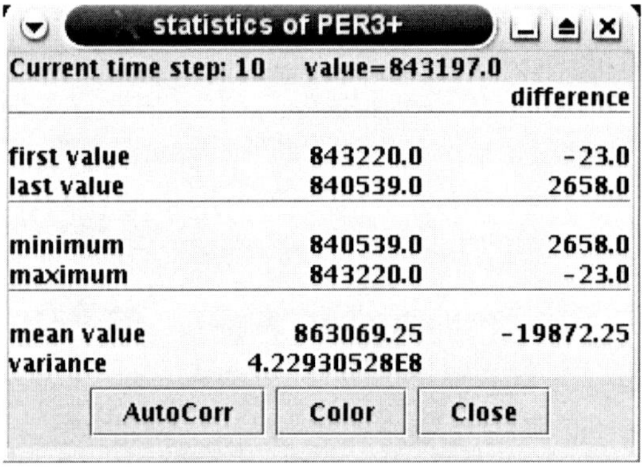

Figure 5.9.: Additional statistical information of a reactant at a certain time step.

used. The number of steps can also be reduced by integrated resampling methods or by setting a minimum concentration change value which filters steps out, in which the concentration change is smaller than this value. This filtering decrease the number of frustums and the response time of the application.

The last feature, a semi-transparent cutting plane, is used to highlight a single time step (that is also integrated in Dwyer et al. [151]). This time step and the corresponding concentration values are also shown in the 2D-View which will be introduced in the next section.

5.3.2. 2D-View

The 2D-View is reused from the SimWiz module to obtain the topology of the network and to visualise the concentration of one single time step in the shapes of the nodes (see figure 5.10). These shapes are circles whose radius is calculated according to the above mentioned radius of conical frustums.

The user has different interaction options to change both views simultaneously. Firstly, by clicking on a shaped circle or by marking a group of circles in this 2D-View, the 3D-View focuses on these marked nodes and the other nodes can be removed or made translucent by choosing a transparency value between 0 and 100%. Single tubes in the 3D-View can also be removed by double clicking on the corresponding reactant in the 2D-View. Secondly, with the time step option the user can change the time step shown in the 2D-View by typing a new value into the text field or by moving the slider. This change evokes a new 2D-View with the concentration values of the chosen time step and moves the semi-transparent cutting plane to the according coordinates of this time step in the 3D-View. Thus, the user knows where this time step is located in the 3D-View.

This view offers the feature of an animation which changes the 2D view and moves the cutting plane in the 3D-View time step by time step automatically starting with the current time step. The speed is adjustable by the user. Furthermore, when the user moves the mouse over a reactant in the 2D-View the corresponding tube in the 3D-View will be highlighted.

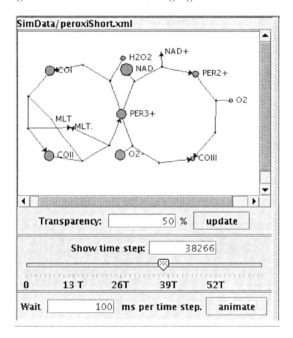

Figure 5.10.: Example of a biochemical reaction network visualised in the 2D-View adopted from SimWiz. The concentration values at a certain time step of a part of the PO reaction are coded in the circle diameters.

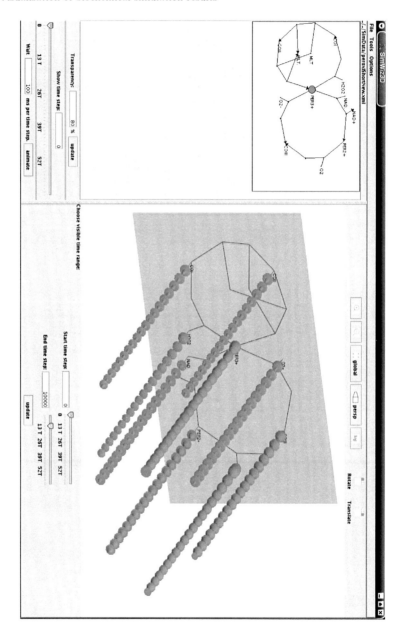

Figure 5.11.: Screen shot of SimWiz3D. The 3D-View shows oscillating time series data.

5.3.3. Settings

In the options menu the user has the choice to make changes to general settings to change colours and to influence the 3D visualisation. Firstly, since the tubes consist of single conical frustum, the user can change the minimum and maximum radius which is useful when only few time series are investigated. Secondly, the number of slices of conical frustums can be increased or decreased. This number influences the smooth appearance of a conical frustum which means the bigger the conical frustum is, the more slices are needed to smooth it out. But the user should bear in mind that the more slices are used, the more time is consumed by the rendering process. Thirdly, the colours of the tubes and the semi-transparent cutting plane can be adapted. Finally, since the resolution of a screen is limited, a maximum number of rendered time steps can be set. This number is used to resample data sets with larger numbers of time steps. The default value is 2000 time steps and the following resampling methods can be chosen:

- nearest time step (dropping steps),

- maximum value of a group,

- mean value of a group.

This resampling method can also be switched off to visualise all time steps or a bigger subset of the data.

5.3.4. Data analysis methods

SimWiz3D offers the standard cross-correlation method and a new method for detecting very small correlations which will be explained in detail in chapter 6. There are two visualisation techniques to visualise the correlation results. Firstly, the correlation matrix can be shown as a table in which the background of the cell of the values are colour coded according to the correlation strength. I defined four types of absolute correlation values between 0 and 1:

- $value < 0.4$ = white (no or small correlation)

- $0.7 < value => 0.4$ = yellow (medium correlation)

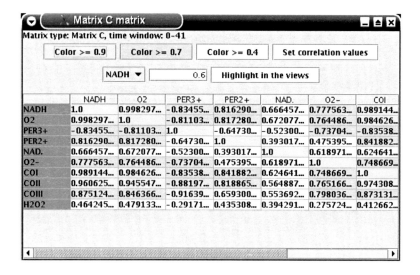

Figure 5.12.: Visualisation of a calculated correlation matrix of ten species as a table.

- $0.9 < value => 0.7 =$ blue (strong correlation)

- $1 < value => 0.9 =$ green (very strong correlation)

These types can also be adapted to other values or colours by the user. Furthermore, the user can set a minimum correlation value and select a specific species in the selection box above the correlation table. All species that are correlated to this species with at least the given correlation value are highlighted in both views. Secondly, the correlation matrix can be represented as grey-scale plot. In this plot the values are mapped to colours between white (1 = perfect correlation) and black (0 = no correlation).

Furthermore, a space-time plot utility is included to visualise the time series data (figure 5.14).

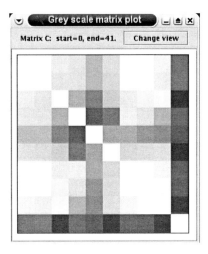

Figure 5.13.: Gray-scale plot of a calculated correlation matrix of ten species.

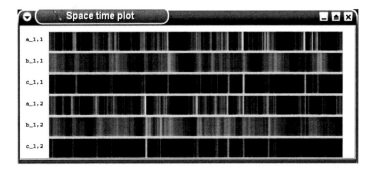

Figure 5.14.: A space-time plot of six species (white - maximum value, black - zero).

5.4. Discussion

SimWiz3D is a tool to visualise biochemical simulation results. It consists of a 3D-View which visualises time series data received from biochemical network simulations and an additional 2D-View to keep track of the network structure. Furthermore, SimWiz3D displays time series data in a better and clearly arranged way than it would be possible with traditional concentration-time plots which enables researchers to analyse numerical data in a comfortable way.

Now I will analyse SimWiz3D by the guidelines given in section 5.2. Shneiderman defined seven tasks visualisation tools should contain:

Overview: The 2D-View gives an overview of the network topology which helps the user to find out relationships between compounds in the network and to compare concentration values at a certain time step. The 3D-View displays an overview about complete time series.

Zoom: The 3D-View contains zoom in and out functions to concentrate on interesting items. For a better navigation rotating and translating is also supported.

Filter: The user has the following filtering techniques to filter uninteresting items out which allow him to control the context of the visualisation [154]:

- selecting single time steps,

- highlighting interesting tubes (colour),

- removing less interesting tubes or making them translucent,

- selecting the visualised time range,

- resampling.

Details-on-demand: This task means that the user gets further information about an item when needed. Therefore, SimWiz3D contains a pop-up window for each compound that shows statistical information about its time series.

Relationships: The 2D-View shows the relationships of the compounds in the pathway. The two views are connected, e.g. by double clicking on a tube in the 3D-View the according node in the 2D-View is highlighted and vice versa. Furthermore, the user can apply correlation methods to find out more relationships.

History: Unfortunately undo functions are not implemented, yet, but several reset functions, e.g. colours or tube properties.

Extraction: Saving images of the visualisation is being developed.

According to Carr's [155] demands, the following user tasks are supported:

- Visualisation of the time series data in a continuous manner. A step-by-step visualisation is also included.

- Combination of network structure and time series data.

- Wide range of user interactions.

- Data analysis (correlation methods).

The volumes of the tubes represent the increasing and decreasing concentration values of the species which fulfils Carr's second demand. Since navigating and zooming do not replace filtering, SimWiz3D contains zooming, rotating, translating and filtering techniques. Carr's last demand is also fulfilled because the 3D-View and the 2D-View are changed synchronously when the user changes settings in one of the views.

Finally, some words about how SimWiz3D considers the human perception. Since colour is a very subjective process and differs individually widely [142], SimWiz3D contains different possibilities to change the colour individually which also prevents the user from overloading his visual system with too much colour information. The 3D perception is supported by shadow, rotation and translation. Furthermore, the user can determine the velocity of the animation. That feature allows the user to adapt the frame rate to his personal visual response time.

In table 5.1 I compare the features of the introduced tools for the visualisation of biochemical simulation results. In summary SimWiz3D offers the most features and more user interacting, focusing, and filtering techniques than all other tools which allows the users to adapt the visualisation to their demands and to explore the data in 2D- or 3D-Views. Moreover, correlation techniques to analyse the data are included and commonly used data formats are supported.

Feature	Talis	SimWiz	Borisjuk	Dwyer	SimWiz3D
Overview	X	X	X	X	X
Zoom	X	X	X	X	X
Rotation				X	X
Filter		X (1)			X (5)
Details-on-Demand					X
Relate	X	X	X	X	X
Step-by-Step	X	X	X		X
Complete time series			X	X	X
Correlations					X
Animation	X	X			X

Table 5.1.: Comparing SimWiz3D with existing tools. This table shows features which are supported by the four existing visualisation tools and the new SimWiz3D for the visualisation of biochemical simulation results.

5.5. Implementation

5.5.1. Architecture

As mentioned in section 3.6 ViPaSi is based on the model - view - controller concept which is shown for this module in figure 5.15. The model saves the reactions and simulation data. It also contains correlation methods and a renderer that creates the 3D elements.

The view consists of classes for the 2D-View, the 3D-View and the visualisation of the correlation results which do not know each other. The 2D visualisation is

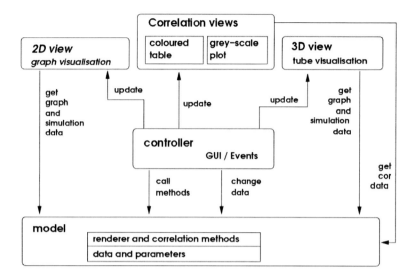

Figure 5.15.: The MVC architecture of SimWiz3D.

part of the SimWiz package and will be reused in the 2D-View of SimWiz3D.

The controller knows each view, registers event listeners, handles user inputs and starts processes to change data and update the views. Additionally, I used several events to communicate between the views and the controller and to synchronise the 2D- and the 3D-View.

5.5.2. The classes in general

The SimWiz3D module contains several classes for each view, the data and the 3D mapping and rendering (see an overview in figure 5.16). As seen in figure 5.16 the graphical user interface of SimWiz3D is based on Java Swing. The classes for the views or the GUI are inherited either from **JFrame** or **JDialog** or **JPanel** of Java Swing (see figure 5.16).

The class **JFrame** is used as parent class for the application class **Sim3DController** and for subwindows showing results (e.g. correlation matrices). I used frames to display results, thus the user can switch between the different visualisations.

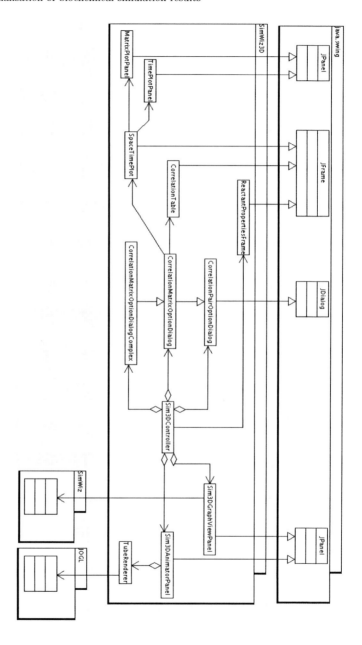

Figure 5.16.: Class diagram of the GUI of SimWiz3D.

Instead, I used the class JDialog as parent class when the application is waiting for user settings before anything can be done. The classes that are part of frames or dialogs are subclasses of the class JPanel.

Each class that has to cope with state changes (events) of components contains a private class that is similar to a mediator design pattern [160]. This private class contains the complete logic to manage events and to coordinate changes of components according to events. Consider as example, if the first slider (start time step) in the 3D-View is changed to a value that is larger than the value of the second slider (end time step), the second slider will be moved automatically to the time step one after the time step in the first slider. Furthermore, the text of the according text fields of this sliders will be set to the chosen values.

5.5.3. The classes in detail

The class Sim3DController coordinates the main window of the application and combines the 3D-View (Sim3DAnimatorPanel) and the 2D-View (Sim3DGraphView-Panel). Furthermore, it defines the menu structure that uses different dialog boxes to change settings, calculate correlations or open files. It accepts and checks the user inputs and shows the according results. If user inputs does not meet the requirements, an error message will be shown containing the requirements and the dialog box for the inputs will be opened again.

Firstly, the user choose a SBML file with layout information and the according parser creates a Vector of GraphicalReaction objects. The class Sim3DController sends an update request to the 2D-View (Sim3DGraphViewPanel) which displays the compounds and reactions in this vector as nodes and edges with the given layout information. Additionally, on the one hand if the SBML file does not contain any layout information, the user can create one with the graph editor in the PathWiz module. On the other hand if the SBML file contains more than one layout, the user can choose which layout should be displayed.

Secondly, the user can choose a simulation data file. All concentration values or particle numbers in this file are saved into numerical arrays, one for each reactant. Thus, the tool needs a longer initial loading time, but allows a faster access for view updates because no further file accesses are needed. Instead of arrays, I could have used vectors or sets but arrays have the following two advantages. Firstly, they are sorted by increasing time. This order is lost in sets. Secondly, since I know the number of time steps, the arrays have a fixed length and can contain primitive data types. For these reasons, they are faster and save memory better than vectors and sets that handle only objects.

The class **ReactantSim3DData** saves the array of the time series data of a compound and the corresponding statistics (e.g. mean value or variance) which are calculated simultaneously when the parser is reading the data. Since the simulation results do not change after the ASCII file was read, this class is immutable which means that the object information cannot be modified after the object was created (read-only) [160]. Therefore, every attribute is private and there are no set-methods which means that no other method as the constructor is able to set the attributes.

After reading the data file the **Sim3DController** class sends an update request to the **Sim3DGraphViewPanel** that integrates the values of the first time step into the current graph representation. Now, since the graph layout and the time series data are loaded it is also possible to initialise the 3D-View (**Sim3DAnimatorPanel**). This panel displays the generated tubes.

The 3D visualisation is done in OpenGL [161] because it is platform independent and offers more flexibility and a wider range on methods than Java3D. I use JOGL [162] for the 3D graphics which is the only Java binding package for OpenGL. The class **TubeRenderer** realises the mapping and rendering of the time series data into OpenGL objects which uses several JOGL classes. The created frustums are cached in display lists that save a precalculated version of them. These precalculated frustums can then be reused multiple times which reduces the rendering time and increases the performance.

Utility classes

Figure 5.17 and 5.18 show several utility classes. The first class **Sim3DParameters** contains the global parameters for the visualisation, e.g. references to the **ReactantSim3DData** objects or the vector of **GraphicalReaction** objects. The second utility class **TubeParameters** consists of parameters that effect only the 3D-View, e.g. tube or plane colour, number of tube slices or stacks. These two classes are designed as Singleton. This design pattern guarantees that there is only one object of this class and therefore each class that needs parameters uses the same set of parameters. Instead, I could have implemented these classes with global attributes that can be accessed and changed by every class. This contains the risk that parameters are changed unintentionally.

The third class Constants is a centralised repository for constant data (e.g. colours) that corresponds to a constant data manager pattern [163]. I designed this class as interface which allows only classes that implement this interface to access the parameters.

The last utility class Sim3DTools collects global methods, e.g. to create Java Swing components (makeButton(), makeLabel() or makeSlider()) which guarantees the similar appearance of components in different panels. The method getGeometry() calculates the width and height of the current graph representation that is used to adapt the panel sizes to the proportion of the layout. If resampling is activated, the method getResamplingSteps() returns an array with the indices of the time steps in the data arrays that shall be displayed in the 3D-View.

Integration of correlation methods

SimWiz3D offers three correlation calculation types:

- pairwise correlation,

- correlation matrix, or

- several correlation matrices and their eigenvalue/eigenvector statistics.

Therefore, I implemented three dialog boxes that build a hierarchy (see figure 5.19). The first class CorrelationPairOptionDialog implements a dialog box for pairwise correlations. The user defines the start and end time step, two compounds and the correlation type. The result is a small information dialog box that shows the correlation coefficient. Since the other two dialog boxes (CorrelationMatrixOptionDialog and CorrelationMatrixOptionDialogComplex) calculate complete correlation matrices, there is no need to choose compounds but they need inputs for the correlation type, start and end time step. Therefore, the two combo boxes in the CorrelationPairOptionDialog class are declared private and all other attributes are protected that allows the usage by the other two dialog boxes.

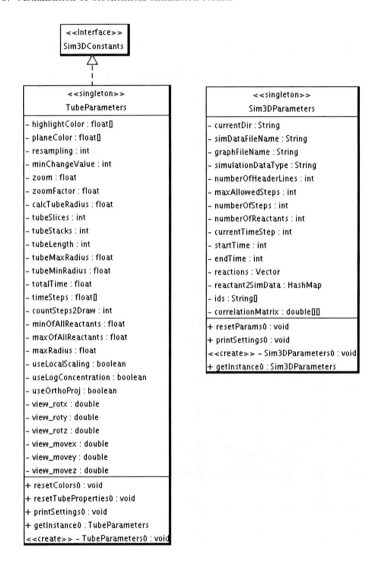

Figure 5.17.: Class diagram of utility classes designed as Singleton.

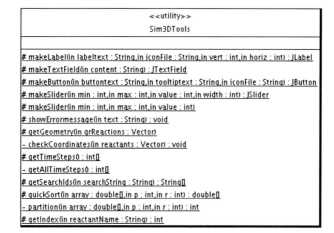

Figure 5.18.: Class diagram of the utility class Sim3DTools that only contains methods.

The class CorrelationMatrixOptionDialog is a subclass of the class Correlation-PairOptionDialog and creates one correlation matrix of the entered time window. It creates an object of the according correlation class which calculates the matrix (see also section 6.5). This matrix can be visualised as grey-scale plot (**SpaceTime-PlotFrame**) or in a table with colour-coding (**CorrelationTableFrame**). I implemented my own table model (**MyTableModel**) and cell renderer (**MyTableCellRenderer**) to realize this colour-coding since it is not possible with the default table model and cell renderer in Java. The correlation values and the background colours for the colour-coding can be changed by the user. The class **ChangeColorValuesDialog** and the class **ColorListener** handle the user changes.

The last dialog box CorrelationMatrixOptionDialogComplex offers the user the possibility to create several matrices and their eigenvalue and eigenvector statistics in one run. It is inherited from the class CorrelationMatrixOptionDialog. Additionally, the class CorrelationMatrixOptionDialogComplex has methods to create the eigenvalue and eigenvector statistics of a number of matrices.

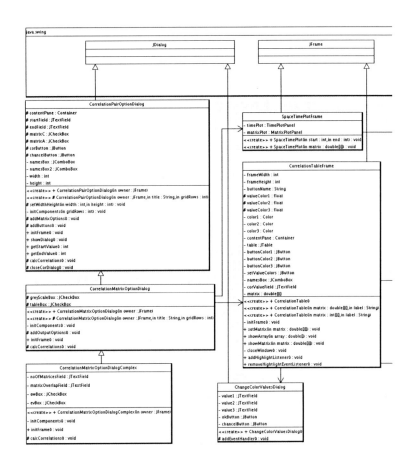

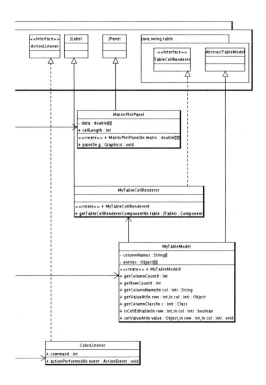

Figure 5.19.: Correlation displays in the SimWiz3D module. The dialog classes are part of the controller and all other classes belong to the view.

Events

I use self-defined events to synchronise the 2D and 3D-View. There are four events (TimeStepChangeEvent, TubeChangeEvent, ViewChangeEvent and HighlightEvent). This event handling is a specialised form of the observer design pattern [160].

The ViewChangeEvent (figure 5.20) is sent by the 2D-View to the 3D-View when the user selected a set of nodes to concentrate on. If the 3D-View gets this event, it will make all nodes, that are not in this set, translucent according to a user defined translucent factor.

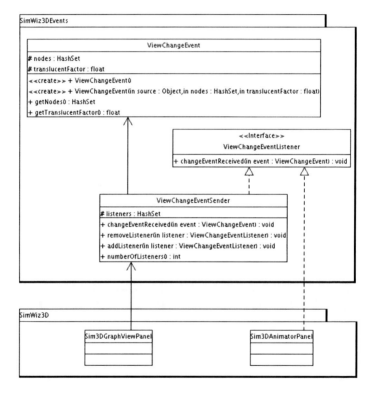

Figure 5.20.: Class diagram of the ViewChangeEvent and its connected classes.

The TimeStepChangeEvent (figure 5.21) informs a view that the user changes a slider in the other view. Therefore, if the start time slider in the 3D-View gets a new value, the 2D-View will update its view to show the new time step. Otherwise, if the user chooses a time step in the 2D-View, the 3D-View will move the cutting plane to this new time step.

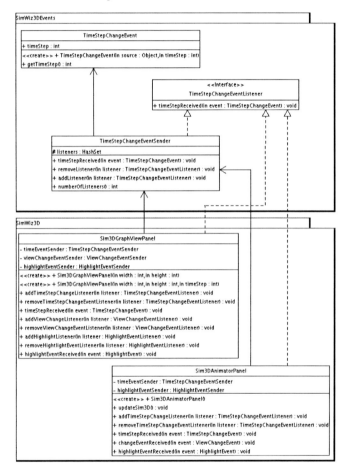

Figure 5.21.: Class diagram of TimeStepChangeEvent and its connected classes.

5. Visualisation of biochemical simulation results

The TubeChangeEvent (figure 5.22) is envoked when the color or the translucent factor of a single tube was changed by the user in the statistics window. According to the new values the 3D renderer creates a new tube which will be displayed in the 3D-View.

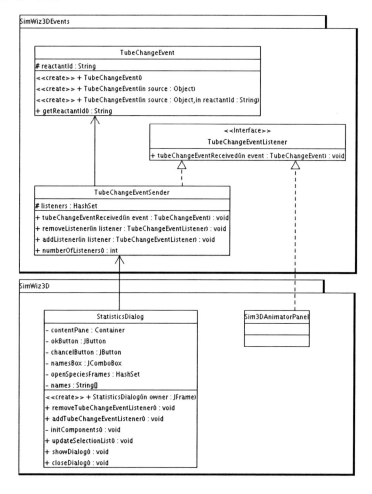

Figure 5.22.: Class diagram of TubeChangeEvent and its connected classes.

The HighlightEvent (figure 5.23) inherits the attributes and methods from the ViewChangeEvent. If the user chooses a compound in the correlation matrix table and a minimum correlation value, the HighlightEvent will be provoked by the CorrelationTable and the 2D-View and the 3D-View will highlight this compound (selectedReactant) and all compounds (selectedNodes) correlated with at least the specified correlation value using a different colour. Furthermore, all tubes of the other compounds (not in nodes) will be made translucent in the 3D-View.

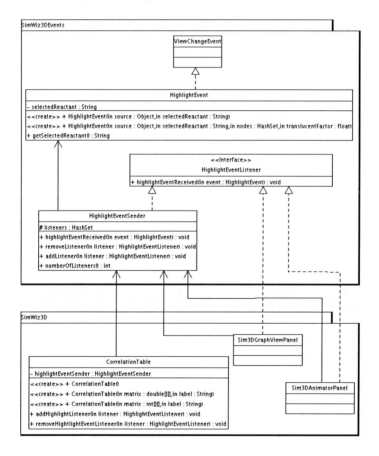

Figure 5.23.: Class diagram of the HighlightEvent and its connected classes.

If the user moves with the mouse over a compound in the 2D-View, the 2D-View will send the **HighlightEvent** to the 3D-View which will highlight the respective tube. The other way around, if the user clicked on a tube in the 3D-View, the 2D-View will get this event from the 3D-View and will highlight the respective compound and all its edges. In these two cases the set of **selectedNodes** and the translucent factor in the **HighlightEvent** will be set to zero and will not be used.

Performance

The most time consuming process is the data mapping when the conical frustums for the 3D-View are created. This process is always invoked when the time range, properties of the frustums, colours, or the underlying time data is changed. The rendering process which is started when the 3D-View is rotated or zoomed using display lists is much faster (see figure 5.24).

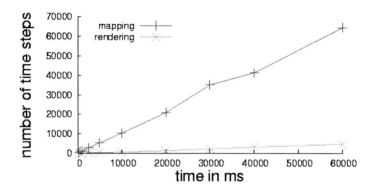

Figure 5.24.: Rendering process vs. mapping process of a system with nine species shown in figure 5.11 (slices: 6, stacks: 1, hardware: AMD Athlon 2000+, 1.6 GHz, nVidia G3 Ti 200, Linux).

Figure 5.24 shows the rendering and mapping times for nine species and a huge amount of time steps. Since the screen resolution is limited, I recommend a much lower number of time steps, e.g. 2000 (default value). Moreover, it is possible to use the integrated resampling methods to get a fast overview of a larger time range.

The mapping time depends mostly on the number of slices and stacks of each frustum. The number of stacks (similar to the latitude) multiplied by the number of slices (longitude) corresponds to the number of polygons that are created for a single frustum. Therefore, the nine tubes in figure 5.24 consists of 108 000 polygons if 2 000 time steps are displayed $(1(stacks)*6(slices)*2000(timesteps)*9(species))$ In this case, the mapping time is $382ms$ and the rendering time is $64ms$ which allows an interactive handling of the 3D-View without delay. Since a frustum represents the change of one time step to the next one, I recommend that the number of stacks should be one. The number of slices depends on the maximum radius and how smooth (round) the tubes should look like (the default value is six). How the variation of the number of slices influences the mapping time is shown in figure 5.25.

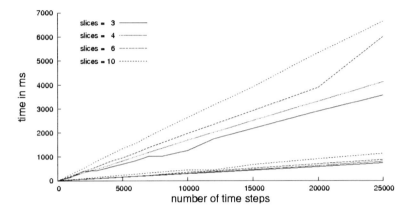

Figure 5.25.: Rendering process vs. mapping process of the system in figure 5.11 with varying number of slices (stacks: 1, hardware: AMD Athlon 2000+, 1.6 GHz, nVidia G3 Ti 200).

As seen in figure 5.25, the higher the number of slices is, the higher the number of polygons is and the slower the mapping time is. If the number of slices is three, the tubes will have a triangular shape and if it is four, a rectangular one. These shapes can be mapped faster, but the thickness of the tubes depends noticeably on the view port of the user.

The frustums are created by the **glyCylinder()** method of OpenGL that is optimised, standardised, and efficient in subdividing the frustums into a minimal number of polygons.

Finally, since time steps with the same concentration values are joined to one conical frustum, the number of polygons is reduced, too.

6. Correlation analysis of simulation data

In the last two chapters I introduced a new dynamic layout algorithm for the graph representation of biochemical pathways and the visualisation package SimWiz3D for the visualisation of biochemical simulation results. Since researchers do not only want to analyse their data visually, I integrated correlation methods into the SimWiz3D package to analyse dependencies.

As I already mentioned in chapter 2.3 there are deterministic and stochastic simulation techniques. These different techniques also produce time series data with different characteristics that need different correlation methods. Furthermore, there are a lot of stochastic and non-linear phenomena that cannot be analysed by conventional correlation methods. Therefore, I will introduce the conventional cross-correlation matrix (matrix **C**) and the new matrix **A** developed in cooperation with Prof. Dr. Markus Müller and Prof. Dr. Gerold Baier, Falcultat de Ciencias, Universidad Autonoma del Estado de Morelos, Cuernavaca, Mexico. Firstly, I will introduce correlation methods in general.

6.1. Correlations in general

The main goal of multivariate data analysis is to simplify and reduce the complexity of the observed system [164]. It helps the user to concentrate on interesting parts and to get new insights into the data. In this chapter I will only consider correlation analysis methods.

The correlation measure answers the question:

Is time series X linearly related to time series Y [165]?

Chen et al. [166] defined three relationships between two variables:

6. Correlation analysis of simulation data

    1. effect Y occurs whenever cause X occurs

    2. effect Y never occurs without cause X having already occurred

    3. effect Y never occurs when cause X has not occurred and effect Y occurs when cause X has occurred

Thus, the correlation describes the statistical dependence of two time series, or in other words it is a *measure of the degree of relationship* [165]. The general correlation coefficient r_{xy} [167] is calculated by

$$r_{xy} = \frac{\sum_{k=1}^{T}(X_k - <X>)(Y_k - <Y>)}{\sqrt{\sum_{k=1}^{T}(X_k - <X>)^2(Y_k - <Y>)^2}} = \frac{s_{xy}}{s_x s_y} \qquad (6.1)$$

where $<X>$ is the mean value of time series X and $<Y>$ is the mean value of time series Y. s_{xy} is the covariance of X and Y and s_x the standard deviation of X and s_y of Y. To get a correlation coefficient between -1 and 1, the data is normalised (mean value = 0, standard deviation = 1). There are three correlation types (see also figure 6.1):

- positive ($r_{xy} > 0$) - the values of Y are increasing/decreasing if the values of X are increasing/decreasing ,

- negative ($r_{xy} < 0$) - the values of Y are increasing if the values of X are decreasing or the other way around,

- no correlation ($r_{xy} = 0$) - no statistical relationship.

The second important correlation method is the autocorrelation given by [168]:

$$r_{lag} = \frac{\sum_{k=1}^{T} X_k X_{t+lag}}{\sum_{k=1}^{T} X_t^2} \qquad (6.2)$$

It detects the relationship of a time series with itself shifted by a certain time lag [168]. The autocorrelation will be only reasonable if the time lag is small compared to the total length T of the observed time series (*lag* $<< T$) [169]. If different lags are calculated the results can be shown in an autocorrelogram (x-coordinate = lag, y-coordinate = autocorrelation coefficient). This autocorrelogram offers conclusions about the interior structure of time series data [168], e.g. the serial dependence [170].

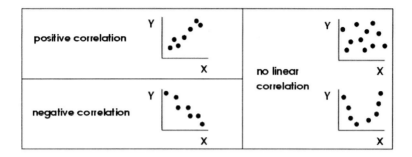

Figure 6.1.: Example plots for positive, negative, and no correlations.

If a multivariate system is analysed, a correlation matrix is calculated that contains a correlation coefficient for each pair of two time series in this system. It is a symmetrical matrix with elements along the diagonal equal to one (perfectly auto-correlated - lag zero). Since observed systems and their complexity are increasing, this equal time correlation matrix C is also used to detect and classify correlations in highly irregular multivariate data sets [171, 172, 173]. Mueller et al. [174] showed that correlations are imprinted in the spectrum of the eigenvectors and eigenvalues of this matrix via level repulsion at the edges of the spectrum. This method allowed a separation of correlations from statistical fluctuations by analysing the smallest and largest eigenvalues and their corresponding eigenvectors, since the central eigenvalues and eigenvectors in the spectrum are dominated by random correlations. This method is very suitable for weakly phase shape correlations. A brief description can be found in section 6.3.1.

Phase shape correlations of complex systems can also be detected by using the synchronisation theory [175]. This method separates the time series in an instantaneous phase and an amplitude at a certain time step via the Hilbert transformation (computation and usage in [175]). Next, the phase differences between two or more time series can be compared to detect relationships or short phase synchronisations. Rosenblum et al. [176] applied this theory to biological simulation results.

Nevertheless, the conventional correlation matrix C and the synchronisation theory are analysing phases, but if the observed system does not have any phases, these methods are not the ideal choice. Therefore, the new matrix $\mathbf{A}$ was developed that will be introduced in section 6.3.2.

In the next section the biochemical test system that was used to evaluate the correlation matrix $\mathbf{C}$ and the new matrix $\mathbf{A}$ will be described.

6.2. Biochemical test system

The test system describes calcium oscillators. Calcium is a very important messenger species in many different cells [177]. In several experiments researchers found out that the frequency of calcium oscillations encode biological information (for a review see [178]) and is usually positively correlated with the agonist concentration. In response to different agonists calcium oscillations show different as well as spatial patterns, e.g. spiking or bursting oscillations [179]. Because of these variety of cell types different models for calcium oscillations have been developed (for a review see [180]).

This test system contains ten cells that are coupled by linear diffusion. The diffusion terms describe the calcium movements between these cells. The concentration changes of the three participating species (calcium and two cell membrane proteins) are described by the following ordinary differential equations that are introduced in detail in [181]:

$$
\begin{aligned}
\frac{dX(t)}{dt} &= k_1 + k_2 X - k_3 \frac{XY}{X + k_4} - k_5 \frac{XZ}{X + k_6}, \\
\frac{dY(t)}{dt} &= k_7 X - k_8 \frac{Y}{Y + k_9}, \\
\frac{dZ(t)}{dt} &= k_{10} X - k_{11} \frac{Z}{Z + k_{12}} = F_Z(X, Z),
\end{aligned}
\tag{6.3}
$$

where Z denotes the calcium in the cell and X and Y correspond to the two membrane proteins responsible for the calcium stimulation. The coupling of the ten cells is described by the following term:

$$
\frac{dZ_i}{dt} = F_Z(X, Z) - D(2Z_i - Z_{i-1} - Z_{i+1}),
\tag{6.4}
$$

where D is the diffusion constant and Z_i the calcium concentration in the ith cell. Since the first and the tenth cell are also coupled, the system has a ring structure.

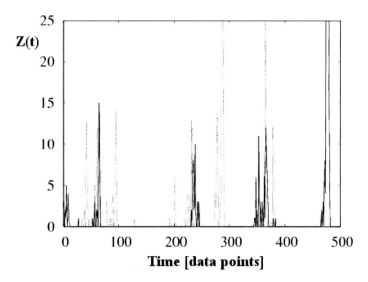

Figure 6.2.: Concentration-time plot of variable 4 and 5 of the test system of ten ring coupled calcium oscillators (parameters: $k_1 = 0.212, k_2 = 3.2, k_3 = 4.88, k_4 = 1.18, k_5 = 1.52, k_6 = 0.19, k_7 = 1.24, k_8 = 32.24, k_9 = 29.09, k_{10} = 13.58, k_{11} = 153, k_{12} = 0.16, D = 0.1$, Volume set to unit).

Since only low number of particles are involved, stochastic simulation is used to generate the time series data (Gillespie algorithm [182]). The resulting time series data shows a bursting characteristic. Two time series of the test system with that bursting characteristic are shown in figure 6.2. This figure shows the three possible cases. Firstly, bursts occur only in one time series (e.g. at 270-290 and 480-490 data points). Secondly, bursts occur in both time series almost simultaneously with approximately the same amplitudes (e.g. at 50-70) and thirdly, bursts occur in both time series almost simultaneously but not with the same amplitude (e.g. at 370-380). The space-time plots in figure 6.3 (done with the SimWiz3D module) show the ten cells uncoupled ((a) $D = 0$) and weakly coupled ((b) $D = 0.1$). Since the time series of this system are characterised by long segments of silence

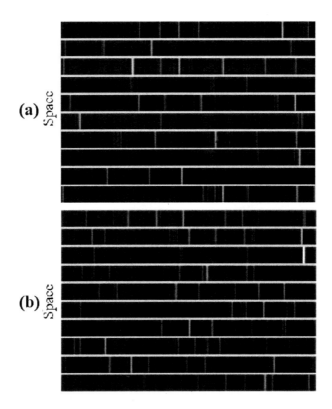

(a) Space

(b) Space

Figure 6.3.: Test system of ten ring coupled calcium oscillators (test system parameters: $k_1 = 0.212, k_2 = 3.2, k_3 = 4.88, k_4 = 1.18, k_5 = 1.52, k_6 = 0.19, k_7 = 1.24, k_8 = 32.24, k_9 = 29.09, k_{10} = 13.58, k_{11} = 153, k_{12} = 0.16$, D $= 0.1$, Volume set to unit). (a) space-time plot of the uncoupled 10 units, (b) space-time plot of the 10 units weakly coupled to a circle with $D = 0.1$.

($Z(t) = 0$) which is aperiodically interrupted by short bursts of irregular shape, these space-time plots contain a lot of black parts for the silence and some grey-coded lines for the bursts. Because of these characteristics you cannot recognise any relationship neither between bursts within one cell, nor between two coupled cells, although there should be correlation due to the coupling of neighbouring cells. Therefore, a very sensitive correlation method is needed that is able to detect even weak correlations in this kind of time series data. Simultaneously the results of these methods must be statistically significant, especially if only short time series data is available which is typical for experimental results in biochemistry.

6.3. Correlation matrices

6.3.1. Matrix C - Standard correlation method

The conventional equal-time correlation matrix analyses a set of multivariate data $Z_i(t_k)$ $(i = 1, \ldots, M)$ over a stationary time window t_k $(k = 1, \ldots, T)$. M denotes to the number of time series and T the number of data points in the time window. T should be considerably larger than M. The data points in the time window t_k are normalised with:

$$\tilde{Z}_i(t_k) = \frac{Z_i(t_k) - <Z_i>}{\sigma_i} \tag{6.5}$$

where $<Z_i>$ is the average amplitude and σ_i the standard deviation, respectively, both calculated for the time window of the length T. The equal time correlation matrix $\mathbf{C}$ [183] can then be constructed as

$$C_{ij} = \frac{1}{T} \sum_k \tilde{Z}_i(t_k)\tilde{Z}_j(t_k) . \tag{6.6}$$

Due to the normalisation (6.5) of the data points the correlation coefficients in the matrix vary between plus and minus one. Negative values denote anti-correlations. The elements along the diagonal in the matrix are equal to one since a time series is perfectly auto-correlated (zero lag). If the correlation matrix is constructed with equation (6.6), due to the normalisation of the time series the sum of the eigenvalues λ_i $i = 1, \ldots M$ is equal to the trace of the matrix

$$\mathrm{Tr}(\mathrm{C}) = \sum_{i=1}^{M} C_{ii} = \sum_{i=1}^{M} \lambda_i. \tag{6.7}$$

Using these equations you can denote the relationships directly between shapes and phases of the analysed time series data. In the case of the test system with a lot of silence and aperiodically bursts a method that quantifies phase-shape-correlations like the correlation matrix formalism [174] may not be the optimal

149

tool of analysis because an unambiguous definition of a phase is not possible. Furthermore, my tests showed that the matrix $\mathbf{C}$ cannot detect a significant difference between an uncoupled and a weakly coupled system. For these reasons, we analyse the occurrence and amplitudes of the bursts instead. Therefore, we developed the new correlation method matrix $\mathbf{A}$ which will be described in the next section and was completely implemented by myself.

6.3.2. Matrix $\mathbf{A}$ - A new correlation method

Since this matrix is supposed to find correlations based on the occurrence and amplitude of bursts in the time series data, we define the following probabilities:

Definition 14 $\mathbf{P}_i$ *is the probability to find a finite calcium concentration at any time point of time series* $Z_i(t)$.

Definition 15 $\mathbf{P}_{ij}$ *is the conditional probability that a finite concentration in time series* Z_j *at any time point occurs when a finite concentration occurs in time series* Z_i.

If both time series are uncorrelated, P_{ij} equals $P_i * P_j$ on the average. The entries A_{ij} of the new matrix $\mathbf{A}$ are calculated by considering the magnitude of the concentration values:

$$A_{ij} = \frac{1}{T_a} \sum_{k=1}^{T} \frac{Z_i(t_k) Z_j(t_k)}{\sqrt{< Z_i^2 >< Z_j^2 >}} \ . \tag{6.8}$$

where $T_a = T \frac{P_i + P_j}{2}$ is the arithmetic average over the number of concentration values of time series Z that are non-zero. $< Z^2 >$ is the average power of a time series. Similar to the matrix $\mathbf{C}$ (6.6) the new matrix $\mathbf{A}$ is symmetric ($A_{ij} = A_{ji}$) and the elements along the diagonal A_{ii} are equal to one. The trace is the sum of the eigenvalues of the new matrix $\mathbf{A}$ and is equal to the number of time series M:

$$\text{Tr}(A) = \sum_{i=1}^{M} A_{ii} = \sum_{i=1}^{M} \Lambda_i = M \tag{6.9}$$

where $\Lambda_i \quad i = 1, ...M$ denote the eigenvalues of the matrix new $\mathbf{A}$. In the uncoupled system state ($D = 0$) the value $A_{ij} = A_i A_j$ is equal to random bursts in time series Z_i and Z_j at the same time steps with

$$A_i = \frac{1}{\sqrt{TP_i}} \sum_{k=1}^{T} \frac{Z_i(t_k)}{\sqrt{<Z_i^2>}} \ . \tag{6.10}$$

In contrast to the matrix $\mathbf{C}$ all non-diagonal entries of the new matrix $\mathbf{A}$ are positive because all concentration values of the test system are equal to or greater than zero (therefore, $A_{ij} > 0$). Furthermore, equation (6.8) includes the case that some finite correlations between *all* units of the system are present, although the cells in the system are uncoupled. In this case the matrix entries are dominated by random correlations. To confirm this statement, further analysis techniques are required that will form a part of the next section.

6.4. Determining random correlations

To determine whether the matrix $\mathbf{C}$ or the new matrix $\mathbf{A}$ are dominated by random correlations the average cross-correlation coefficient c and its standard deviation σ_c are calculated of the non-diagonal matrix elements. Using these three values three matrices for the matrix $\mathbf{C}$ and three for the new matrix $\mathbf{A}$ are created with entries along the diagonal equal to one and all non-diagonal entries equal to:

- c (average cross-correlation coefficient)

$$\mathbf{C}/A_{av} = \begin{pmatrix} 1 & & \\ c & 1 & c \\ & & 1 \end{pmatrix} \tag{6.11}$$

- $c + \sigma_c$

$$\mathbf{C}/A_+ = \begin{pmatrix} 1 & & \\ c + \frac{\sigma_c}{2} & 1 & c + \frac{\sigma_c}{2} \\ & & 1 \end{pmatrix} \tag{6.12}$$

- $c - \sigma_c$ are constructed.

$$\mathbf{C}/A_- = \begin{pmatrix} 1 & & \\ c - \frac{\sigma_c}{2} & 1 & c - \frac{\sigma_c}{2} \\ & & 1 \end{pmatrix} \tag{6.13}$$

6. *Correlation analysis of simulation data*

The last two matrices (6.12 and 6.13) provide a measure for the statistical error. By induction you can prove that the characteristic polynomial of such matrices is given by

$$P_M(\lambda) = (1 - \lambda - B)^{(M-1)}(1 - \lambda + (M - 1)B) \tag{6.14}$$

where B is the value of the non-diagonal elements of the three matrices. Hence, each matrix has one large eigenvalue, namely

$$\Lambda_l = 1 + (M - 1)B \tag{6.15}$$

and $(M - 1)$ degenerate small eigenvalues

$$\lambda_s = 1 - B. \tag{6.16}$$

These eigenvalues and the statistical error as a function of the coupling strength D over a time window of $T = 150000$ time steps of the test system are shown in figure 6.4. As seen in figure 6.4, in contrast to the new matrix $\mathbf{A}$ the largest eigenvalue Λ_l and the smallest eigenvalue λ_s of matrix $\mathbf{C}_{av}$ are equal to one because in the uncoupled case $(D = 0)$ C_{ij} is equal to zero on the average. Since A_i of equation (6.10) is always greater than zero, non-diagonal entries are also always greater than zero for the time series data of the test system, even if $D = 0$ (because of random coincidences). However, since the difference

$$\Delta_c(D) = \lambda_l(D) - \lambda_s(D) \tag{6.17}$$

of the largest and all other eigenvalues is increasing with increasing coupling, the following correlation measure for the whole system can be defined

$$Cor(D) = \frac{\Delta(D) - \Delta(D = 0)}{M - \Delta(D = 0)} \tag{6.18}$$

where $\Delta(D)$ denotes $\Delta_a(D)$ or $\Delta_c(D)$ respectively. Theoretically, $Cor(D)$ has a range between zero (no correlation) and one (perfectly correlated), but due to the stochastic nature of the time series data, $Cor(D)$ will never reach its maximum.

Figure 6.5 compares the values of the correlation measure for the matrix $\mathbf{C}$ (dashed line) and the matrix new $\mathbf{A}$ (solid line). On the one hand, the correlation measure of the new matrix $\mathbf{A}$ gets always for $D > 0$ higher correlation values and

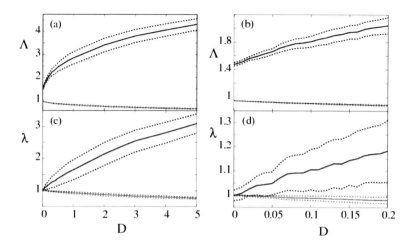

Figure 6.4.: Eigenvalue comparison of matrix **C** and the new matrix **A** as function of the coupling constant D. (a) Eigenvalues of $\mathbf{A}_{av}$ (solid line) and $\mathbf{A}_\pm$ (dashed line), (b) amplification of figure (a) for the range $D = 0$ to 0.2, (c) eigenvalues of $\mathbf{C}_{av}$ (solid line) and $\mathbf{C}_\pm$ (dashed line), (d) amplification of figure (c) for the range $D = 0$ to 0.2.

increases especially for small couplings more rapidly than the correlation measure for the matrix **C** (e.g. $D = 0.2$ Cor_A is three times larger than Cor_C). On the other hand the larger the coupling, the slower the differences are which shows that matrix **C** is also suitable for such data if the coupling is high enough. This approximation is expected because the stronger the coupling is, the stronger the correlations between the ten cells of the test system will be. This comparison proves that the matrix **A** is more sensitive than the matrix **C** for small couplings.

The statistical significance of $C(D)$ is estimated by

$$S(D) = \exp\left[-\frac{2(\Delta(D) - \Delta(D = 0))}{\Sigma(D)}\right] \qquad (6.19)$$

where $\Sigma(D) = (\sigma_l(D) + \sigma_s(D))$ is the sum of the error bars of the large and the small eigenvalue $(\sigma_l(D) = |\lambda_{l+} - \lambda_{l-}|$ and $\sigma_s(D) = |\lambda_{s+} - \lambda_{s-}|)$. The exponential function in (6.19) is used for a clearer differentiation between statistical significance and random correlations.

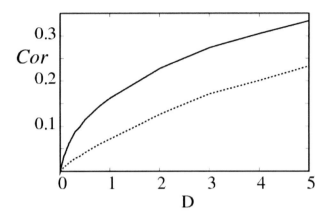

Figure 6.5.: Correlation measure Cor calculated from the new matrix **A** (solid line) and the matrix **C** (dashed line) as a function of the coupling strength D. A window of 150 000 data points was used to construct the matrices.

If the statistical error $\Sigma(D)/2$ is smaller than the average separation of the eigenvalues $\Delta(D) - \Delta(D = 0)$ the value of $S(D)$ will be smaller than $1/e$, otherwise it takes values between zero and $1/e$. Figure 6.6 shows $S(D)$ as function of the coupling strength D for the matrix **C** (6.6) and the new matrix **A** (6.8) and the value $1/e$. At $D = 0.2$ the value of $S(D)$ for the new matrix **A** is about two magnitudes smaller than the calculated value for the matrix **C** which shows the higher significance of the new matrix **A** for small couplings.

To show that the new method is also suitable for rather short time series data, you see in figure 6.7 the correlation measure $Cor(D)$ and significance measure $S(D)$ for matrix **C** and the new matrix **A** calculated for $T = 1000$ data points (which corresponds to approximately 20 bursts in the test system). The curves are the average result over 100 matrices.

The statistical significance $S(D)$ (equation 6.19) for the new matrix **A** crosses already at $D \approx 0.05$ the critical line of value e^{-1}, whereas $S(D)$ for matrix **C** crosses this line at $D \approx 0.45$. This fact is remarkable for the new matrix **A** and shows the improvement of correlation analysis when weakly correlations exist in time series data with bursting characteristics.

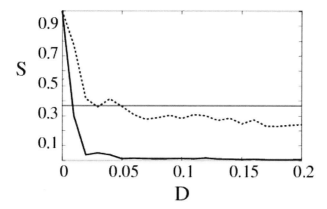

Figure 6.6.: This figure shows the statistical significance $S(D)$ as function of the coupling D for the new matrix $\mathbf{A}$ (solid line) and the matrix $\mathbf{C}$ (dashed line). As comparison the value e^{-1} is drawn as a horizontal solid line. A time window of 150 000 sampling points was used to construct the matrices.

In summary, the new matrix $\mathbf{A}$ is suitable for the detection of correlation in time series data with bursting characteristics and has a higher significance for weak correlations than the conventional matrix $\mathbf{C}$. Furthermore, the method is also applicable for rather short time series data.

Additionally, statistical analysis techniques for the eigenvectors of the matrices are implemented. These eigenvectors show differences when e.g. instead of ten coupled cells in the test system only two out of ten cells are coupled. Thus, if for example only two of the ten cells in the test system are coupled ($M = 10$), eigenvectors corresponding to the largest and smallest eigenvalues show significantly larger values of those components that correspond to the two correlated cells (e.g. figure 6.8).

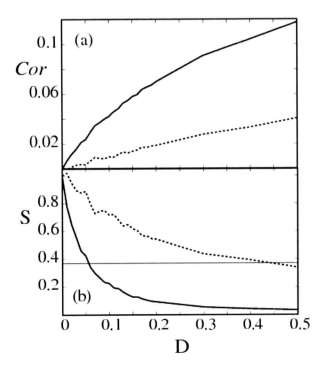

Figure 6.7.: This figure also shows the statistical significance $S(D)$ as function of the coupling. but the time window only has a length of 1000 data points. Matrix $\mathbf{A}$ (solid line), matrix $\mathbf{C}$ (dashed line) and value e^{-1}(solid line).

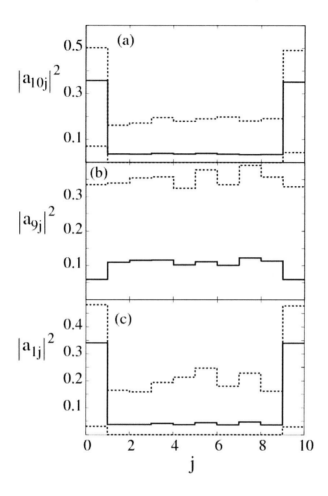

Figure 6.8.: Eigenvector components from A calculated for 500 matrices over a data segment of 15 000 sampling points per matrix when only two from ten units of the test system are coupled with D = 0.1. (a) Components of the tenth eigenvector corresponding to the largest eigenvalue 10, (b) components of the ninth eigenvector, and (c) components of the first one. The difference between the first and the tenth eigenvector to all other eigenvectors (all similar to the ninth) shows that the first and the tenth cell are coupled. Solid line: average, dashed lines: 95% confidence levels

6.5. Implementation

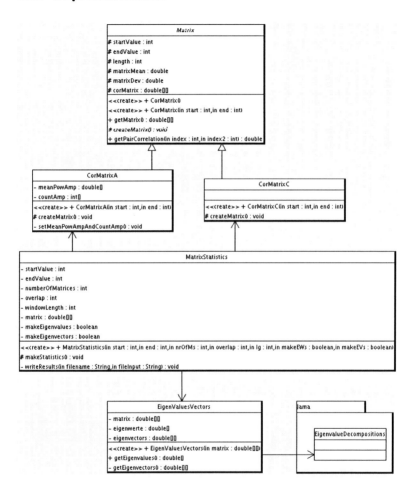

Figure 6.9.: Class diagram of the classes to calculate the correlation matrices described above.

Figure 6.9 displays the class diagram of the classes for the above mentioned correlation methods. I created a template class **Matrix** which contains the common parts of matrices in general and the abstract method **createMatrix()**. This method is implemented by the two subclasses **CorMatrixC** and **CorMatrixA** according to the methodology description above. I did not reuse an existing matrix implementation because if I create my own matrix class, there will be no overhead of attributes or methods that are not needed in this application.

The third class calculates the eigenvalue and eigenvectors of one matrix by using the open-source Java package Jama (**J A**va **MA**trix package) [184]. The class **MatrixStatistics** creates a number of matrices and the statistic of the eigenvalues and eigenvectors of these matrices. The results are saved into a file that can be visualised e.g with gnuplot [185].

These classes are part of the model part in the SimWiz3D module that offers several dialog boxes in the controller part. These dialog boxes accept user settings and start the correlation calculations. The results are visualised in a table with colour-coding or in a grey-scale plot (view part).

6. Correlation analysis of simulation data

7. Summary

Biochemists are interested in processes of living cells. Since cells have an unlimited variety of forms and functions, these processes are very complex. They consist of reactions in which species in cells are used to produce or rebuilt other species. If reactions share species, these connected reactions are called biochemical pathways/networks. Biochemists use more and more modelling and simulation techniques on the computer to analyse and expand this knowledge.

In this thesis I introduced the package **ViPaSi** (**Vi**sualisation of biochemical **Pa**thways and their **Si**mulation results) that helps researchers to create models and to interpret their simulation results by using new visualisation techniques and correlation analysis methods.

In the modelling process a mathematical model is generated that describes the concentration or number of particle changes of species. To ease the modelling process a graph representation is used to get an overview of the relationships between the species in a pathway. In such a graph nodes represent the species and edges represent the reactions. Mostly, the layout of these graphs is handmade by researchers which costs a lot of time. Therefore, some dynamic layout algorithms for biochemical pathways have been developed, but mostly for database usage. In this case they are not suitable for modelling and simulation tools because additional information from databases that is used during the layout process is not available. Furthermore, since the existing layout algorithms have drawbacks according to finding cycles with joined nodes or finding small cycles that represent important recycling processes or short cuts, a new layout algorithm was developed in this thesis.

I described this algorithm in detail and the comparison with existing algorithms shows that the above mentioned drawbacks are solved by the new dynamic layout algorithm for biochemical pathways. Since the resulting layout should be similar to the graph representations biochemists are used to, the new layout algorithm considers a lot of biological conventions and searches for the smallest instead of the longest cycle independently from edge directions as well as finds cycles that share nodes. Furthermore, it reduces edge crossings by splitting nodes and is independent from additional information. In order to share layout information between different tools, I supported the development of an exchange format for layout information for biochemical pathways.

After modelling the pathway the simulation solves the mathematical model and creates a high amount of multivariate time series data. Researchers want to analyse and interpret this time series data to get more insights into the biochemical processes in living cells. Therefore, suitable visualisation techniques are necessary. Most simulation tools have in common that they offer only a concentration-time-plot visualisation of this data, but these plots are not suitable for the analyses of a higher number of species. Only three of these tools have further visualisation techniques but these are only suitable for special purposes, e.g. for the visualisation of diffusion processes. Therefore, the new visualisation tool SimWiz3D was presented. It offers an alternative visualisation for multivariate time series data in general as well as a lot of user interactions. It combines the graph representation of the model with time series data in different views. The 3D view visualises the time series data in a continuous manner and the 2D view shows single time steps.

Moreover, I integrated cross-correlation techniques that help to interpret the time series data. Since I found out in my correlation studies that the standard cross-correlation method is not sensitive enough for bursting data with small correlations, I supported the development of the new correlation method matrix **A** and implemented it.

Summarised, ViPaSi is a platform independent tool written in Java. It consists of three modules. Firstly, the PathWiz module contains a graph editor and the new dynamic layout algorithm for the visualisation of biochemical pathways. Secondly, SimWiz offers a step-by-step visualisation of multivariate time series data. Thirdly, SimWiz3D visualises the time series data in a continuous manner. All modules can also be used as stand-alone applications or as add-on for every modelling and simulation tool.

7.1. Outlook

Since living cells consist of compartments, it would be desirable that the new dynamic layout algorithm for biochemical pathways would take compartments into account. That could be done by creating one graph for each compartment, then each graph would be layouted by the new layout algorithm. Finally, the graphs would be reassembled to one complete graph.

SimWiz3D could be supplemented by further correlation or time series analysis methods depending on what users need. Furthermore, the results of network analysis (e.g. flux modes) or complexity reduction algorithms could be integrated into the visualisation.

7. Summary

A. Sequence diagrams

This chapter contains sequence diagrams of the existing dynamic layout algorithm by Becker et al. and the newly developed algorithm which shows the differences of the both algorithms in a clearer way. The first sequence diagram (figure A.1) shows the algorithm of Becker et al.. The following three diagrams (figures A.2, A.3, A.4) describe the new algorithm.

A. *Sequence diagrams*

A.1. Becker et al.

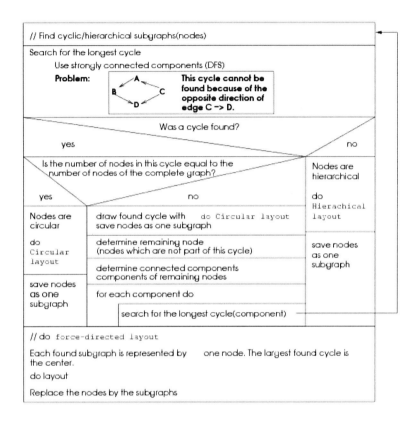

Figure A.1.: Sequence diagram of the layout algorithm of Becker et al..

A.2. New dynamic layout algorithm

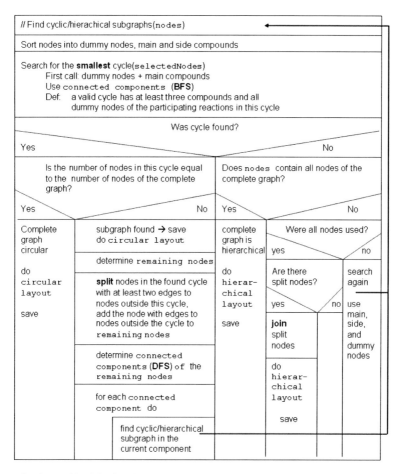

Figure A.2.: Sequence diagram of the new layout algorithm (1/3) displaying the method findSubgraph() (see figure 4.15).

// Join split nodes in cycles (cycles share one node)		
Sort split nodes into main and side compounds		
// Join main compounds before side compounds		
For each split node (node n1 (first cycle) and n2 (second cycle))		
	Rotate both cycles to each other at these nodes	
	Move the second cycle at n2 to the position of n1 in the first cycle	
	Remove node n2 from the graph	
	Move all edges of node n2 to node n1	

Is the number of found circular and hierarchical subgraphs > 0?

Yes — No

// Find new cycles that share more than one node		
For each cycle		

Are there two edges to a hierarchical subgraph?

Yes — No

| Find path1 between the nodes of these edges in the cycle |
| Find path2 between the nodes of these edges in the hierarchical subgraph |
| Build a new cycle with nodes of path1 and path2 |
| Join the current cycle and the new cycle at path1 |

Is the sum of found circular subgraphs + hierarchical subgraphs > 1?

Yes — No

| // Reassemble subgraphs |
| Place the cycle with the highest number of edges to other subgraphs as central subgraph |
| do force-directed |

Figure A.3.: Sequence diagram of the new layout algorithm (2/3).

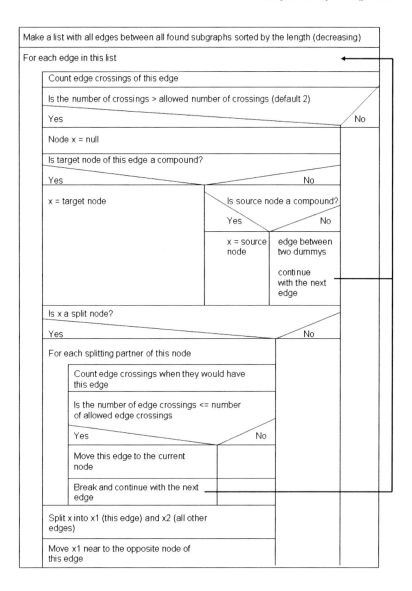

Figure A.4.: Sequence diagram of the new layout algorithm (3/3).

A. Sequence diagrams

B. Publications

Katja Wegner and Ursula Kummer, "A new dynamical layout algorithm for complex biochemical reaction networks", *BMC Bioinformatics*, 6:212, 2005 (http://www.biomedcentral.com/1471-2105/6/212).

Katja Wegner, "SimWiz3D - Visualising biochemical simulation results", in *Proceedings of MediViz 2005. IEEE Computer Society*, 2005, pp. 77-82.

Markus Müller, Katja Wegner, Ursula Kummer and Gerold Baier, "The quantification of cross-correlations in complex spatio-temporal systems", *Physical Reviews E*, 73:046106, 2006.

Ralph Gauges, Ursula Rost, Sven Sahle and Katja Wegner, "A Model Diagram Layout Extension for SBML", *Bioinformatics*, in Press.

B. Publications

Bibliography

[1] M. W. Davidson, http://micro.magnet.fsu.edu/cells/animals/animal-model.html, 2005.

[2] H. Kitano, "A graphical notation for biochemical networks," *BIOSILICO*, vol. 1, no. 5, pp. 169–176, 2003.

[3] G. D. Battista, P. Eades, R. Tamassia, and I. G. Tollis, *Graph Drawing: Algorithms for the Visualization of Graphs*. Prentice Hall, New Jersey, 1999.

[4] Wikipedia, http://de.wikipedia.org/wiki/Hauptseite, 2005.

[5] T. Reiß, *Systems Biology*. Bonn: Federal Ministry of Education and Research (BMBF), 2002.

[6] K. Takahashi, K. Yugi, K. Hashimoto, Y. Yamada, C. J. F. Pickett, and M. Tomita, "Computational Challenges in Cell Simulation," *IEEE Intelligent systems*, 20026471.

[7] D. Endy and R. Brent, "Modelling cellular behaviour," *Nature*, vol. 409, pp. 391–395, 2001.

[8] E. O. Voit, *Computational Analysis of Biochemical System*. Cambridge: Cambridge University Press, 2000.

[9] F. J. Brandenburg, B. Gruber, M. Himsolt, and F. Schreiber, "Automatische Visualisierung biochemischer Information," in *Workshop Molekulare Bioinformatik, GI Jahrestagung*, 1998, pp. 24–38.

[10] G. Michal, *Biochemical Pathways (Poster)*. Boehringer Mannheim GmbH - Biochemica, 1993.

[11] ——, *Biochemical Pathways.* Spektrum Akademischer Verlag, 1999.

[12] M. Kanehisa, "Toward pathway engineering: a new database of genetic and molecular pathways," *Science and Technology Japan*, vol. 59, pp. 34–38, 1996.

[13] B. Aleman-Meza, Y. Yu, H.-B. Schüttler, J. Arnold, and T. R. Taha, "KINSOLVER: A simulator for computing large ensembles of biochemical and gene regulatory networks," lsdis.cs.uga.edu/~aleman/kinsolver/paper25.pdf, 2005.

[14] B. Palsson, "The challenge of in silico biology," *Nature biotechnology*, vol. 18, pp. 1147–1150, 2000.

[15] H. de Jong, "Modeling and Simulation of genetic regulatory systems: a literature review," *Journal of Computational Biology*, vol. 9, no. 1, pp. 67–103, 2002.

[16] R. Alur, C. Belta, F. Ivancic, V. Kumar, H. Rubin, J. Schug, O. Sokolsky, and J. Webb, "Visual Programming for Modeling and Simulation of Biomolecular regulatory Networks," in *High Performance Computing - HiPC 2002: 9th International Conference Bangalore, India, December 18-21, 2002.*, 2002, pp. 702 – 712.

[17] K. Vasudeva and U. S. Bhalla, "Adaptive stochastic-deterministic chemical kinetic simulations," *Bioinformatics*, vol. 20, no. 1, pp. 78–84, 2004.

[18] K. Ichikawa, "A-Cell: graphical user interface for the construction of biochemical reaction models," *Bioinformatics*, vol. 17, no. 5, pp. 483–484, 2001.

[19] A-Cell, http://www.his.kanazawa-it.ac.jp/~ichikawa/A-Cell/A-Cell_J.htm, 2005.

[20] D. Adalsteinsson, D. McMillen, and T. C. Elson, "Biochemical Network Stochastic Simulator (BioNetS): software for stochastic modeling of biochemical networks," *BMC Bioinformatics*, 2004. [Online]. Available: http://www.biomedcentral.com/1471-2105/5/24

[21] BioNetS, http://x.amath.unc.edu:16080/BioNetS/JavaVersion.html, 2005.

[22] P. Dhar, T. C. Meng, S. Somani, L. Ye, A. Sairam, M. Chitre, Z. Hao, and K. Sakharkar, "Cellware - a multi-algorithmic software for computational systems biology," *Bioinformatics*, vol. 20, no. 8, pp. 1319–1321, 2004.

[23] Cellware, http://www.bii.a-star.edu.sg/research/sbg/cellware/index.asp, 2005.

[24] Copasi, http://www.copasi.org, 2005.

[25] I. Goryanin, T. C. Hodgman, and E. Selkov, "Mathematical simulation and analysis of cellular metabolism and regulation," *Bioinformatics*, vol. 15, no. 9, pp. 749–758, 1999.

[26] S. Ramsey, D. Orrell, and H. Bolouri, "DIZZY: Stochastical simulation of large-scale genetic regulation networks," *JBCB Preprint*, 2005. [Online]. Available: http://www.speakeasy.net/~sramsey/JBCB_2005.pdf

[27] Dizzy, http://labs.systemsbiology.net/bolouri/software/Dizzy/, 2005.

[28] L. You, A. Hoonlor, and J. Yin, "Modeling biological systems using Dynetica: a simulator of dynamic networks," *Bioinformatics*, vol. 19, no. 3, pp. 435–436, 2003.

[29] Dynetica, http://www.duke.edu/~you/Dynetica_page.htm, 2005.

[30] M. Tomita, K. Hashimoto, K. Takahashi, T. Shimizu, Y. Matsuzaki, F. Miyoshi, K. Saito, S. Tanida, K. Yugi, J. C. Venter, and C. Hutchison, "E-CELL: Software environment for whole cell simulation," *Bioinformatics*, vol. 15, no. 1, pp. 72–84, 1999.

[31] Ecell, http://ecell.sourceforge.net/, 2005.

[32] GENESIS, http://www.genesis-sim.org/GENESIS/, 2005.

[33] M. Vass, N. Allen, C. A. Shaffer, N. Ramakrishnan, L. T. Watson, and J. J. Tyson, "The JigCell Model Builder and Run Manager," *Bioinformatics*, vol. 20, no. 18, pp. 3680–3681, 2004.

[34] JigCell, http://jigcell.biol.vt.edu/, 2005.

[35] J. A. Miller, A. F. Seila, and X. Xiang, "The JSIM Web-Based Simulation Environment," *Future Generation Computer Systems (FGCS), Special Issue on Web-Based Modeling and Simulation*, vol. 17, no. 2, pp. 119–133, 2000.

[36] JSim, http://nsr.bioeng.washington.edu/PLN/Members/butterw-/JSIMDOC1.6/JSim_Home.stx/view, 2005.

[37] B. G. Olivier and J. L. Snoep, "Web-based kinetic modelling using JWS Online," *Bioinformatics*, vol. 20, p. 13, 2004.

[38] J. Hattne, D. Fange, and J. Elf, "Stochastic reaction-diffusion simulation with MesoRD," *Bioinformatics*, vol. 21, no. 12, pp. 2923–2924, 2005.

[39] MesoRD, http://mesord.sourceforge.net/, 2005.

[40] T. Zhu, C. Phalakornkule, S. Ghosh, I. E. Grossmann, R. R. Koepsel, M. M. Ataai, and M. M. Domach, "A Metabolic Network Analysis and NMR Experimental Design Program with User Interface-Driven Model Construction," *Metabolic Engineering*, vol. 5, pp. 74–85, 2003.

[41] Metabologica, http://www.metabologica.com/, 2005.

[42] L. Lok and R. Brent, "Automatic generation of cellular reaction networks with moleculizer 1.0," *Nature Biotechnology*, vol. 23, pp. 131–136, 2005.

[43] Moleculizer, http://www.molsci.org/~lok/moleculizer/, 2005.

[44] M. Hucka, A. Finney, H. M. Sauro, H. Bolouri, J. Doyle, and H. Kitano, "The ERATO Systems Biology Workbench: enabling interaction and exchange between software tools for computational biology," in *Pacific Symposium on Biocomputing*, 2002, pp. 450–461.

[45] SBW, http://sbw.sourceforge.net/, 2005.

[46] H. G. Holzhutter and A. Colosimo, "SIMFIT: a microcomputer software-toolkit for modelistic studies in biochemistry," *Bioinformatics*, vol. 6, pp. 23–28, 1990.

[47] SimFit, http://www.simfit.man.ac.uk/, 2005.

[48] Simpathica, http://bioinformatics.nyu.edu/Projects/Simpathica/, 2005.

[49] M. Ander, P. Beltrao, B. D. Ventura, J. Ferkinghoff-Borg, M. Foglierini, A. Kaplan, C. Lemerle, I. Tomas-Oliviera, and L. Serrano, "SmartCell, a framework to simulate cellular processes that combined stochastic approximation with diffusion and localisation: analysis of simple networks," *IEE Systems Biology*, vol. 1, no. 1, pp. 129–138, 2004.

[50] SmartCell, http://www.embl.de/ExternalInfo/serrano/smartcell/, 2005.

[51] Stochastirator, http://opnsrcbio.molsci.org/stochastirator/stoch−main.html, 2005.

[52] N. L. Novère and T. S. Shimizu, "Stochsim: modelling of stochastic biomolecular processes," *Bioinformatics*, vol. 17, no. 6, pp. 575–576, 2001.

[53] Vcell, http://www.vcell.org/, 2005.

[54] XPPAUT, http://www.math.pitt.edu/∼bard/xpp/xpp.html, 2005.

[55] R. A. Earnshaw and N. Wiseman, *An Introductory Guide to Scientific Visualization.* Springer-Verlag, 1992.

[56] M. Hucka, A. Finney, H. M. Sauro, H. Bolouri, J. C. Doyle, H. Kitano, A. P. Arkin, B. J. Bornstein, D. Bray, A. Cornish-Bowden, A. A. Cuellar, S. Dronov, E. D. Gilles, M. Ginkel, V. Gor, I. I. Goryanin, W. J. Hedley, T. C. Hodgman, J. H. Hofmeyr, P. J. Hunter, N. S. Juty, J. L. Kasberger, A. Kremling, U. Kummer, N. Le Novere, L. M. Loew, D. Lucio, P. Mendes, E. Minch, E. D. Mjolsness, Y. Nakayama, M. R. Nelson, P. F. Nielsen, T. Sakurada, J. C. Schaff, B. E. Shapiro, T. S. Shimizu, H. D. Spence, J. Stelling, K. Takahashi, M. Tomita, J. Wagner, and J. Wang, "The systems biology markup language (sbml): a medium for representation and exchange of biochemical network models," *Bioinformatics*, vol. 1, pp. 524–531, 2003.

[57] SBML, http://www.sbml.org, 2005.

[58] R. Gauges, S. Sahle, U. Rost, and K. Wegner, "SBML Layout Extension," http://projects.villa-bosch.de/bcb/sbml/, 2005.

[59] A. M. Finney and M. Hucka, "Systems biology markup language: Level 2 and beyond," *Biochemical Society Transactions*, no. 31, pp. 1472–1473, 2003.

[60] B. J. Bornstein, "LibSBML API reference manual," http://www.sbml.org/-software/libsbml, 2004.

[61] S. Download, http://atlas/bcb/software/, 2005.

[62] P. D. Karp and S. M. Paley, "Representations of Metabolic Knowledge: Pathways," in *Second International Conference on Intelligent Systems for Molecular Biology; Mento Park, CA*, R. Altman, D. Brutlag, P. D. Karp, R. Lathrop, and D. Searls, Eds.   AAAI Press, 1993, pp. 225–238.

[63] JDesigner, http://www.cds.caltech.edu/~hsauro/JDesigner.htm, 2005.

[64] L. F. Olsen, "Simulations of Oscillations of NAD(P)H and Reactive Oxygen Species in Neutrophilic Leukocytes," in *2nd Workshop on Computation of Biochemical Pathways and Genetic Networks*, R. Gauges, C. van Gend, and U. Kummer, Eds., 2001, pp. 89–99.

[65] F. Buschmann, R. Meunier, H. Rohnert, P. Sommerlad, and M. Stal, *Pattern-orientierte Software-Archtektur*.   Bonn, Paris: Addison Wesley Longman Verlag GmbH, 1998.

[66] I. Herman, G. Melancon, and M. S. Marshall, "Graph Visualization and Navigation in Information Visualization: a Survey," *IEEE CS Society*, 2000.

[67] C. Berge, *Graphs and Hypergraphs*.   North-Holland Publishing Company, 1976.

[68] T. H. Cormen, C. E. Leiserson, and R. L. Rivest, *Introduction to Algorithms*. London: MIT Press, 1990.

[69] J. N. Warfield, "Crossing Theory and Hierarchy Mapping," *IEEE Trans. Sys., Man, and Cybernetics*, vol. 7, no. 7, pp. 505–523, 1977.

[70] M. J. Carpano, "Automatic display of hierarchized graphs for computer aided decision analysis," *IEEE Transactions on Systems, Man and Cybernetics*, vol. 10, no. 11, pp. 705–715, 1980.

[71] K. Sugiyama, S. Tagawa, and M. Toda, "Methods for Visual Understanding of Hierarchical System Structures," *IEEE Transactions on Systems, Man and Cybernetics SMC*, vol. 11, pp. 109–125, 1981.

[72] P. Eades and K. Sugiyama, "How to draw a directed graph," *Journal of Information Processing*, vol. 13, no. 4, pp. 424–437, 1990.

[73] M. Eiglsperger, M. Siebenhaller, and M. Kaufmann, "An Efficient Implementation of Sugiyama's Algorithm for Layered Graph Drawing." in *Graph Drawing*, 2004, pp. 155–166.

[74] P. Healy and N. S. Nikolov, "How to Layer a Directed Acyclic Graph," in *Graph Drawing*, 2001, pp. 16–30.

[75] G. Sander, "Layout of Directed Hypergraphs with Orthogonal Hyperedges," in *Graph Drawing*, 2003, pp. 381–386.

[76] M. Forster, "A Fast and Simple Heuristic for Constrained Two-Level Crossing Reduction," in *Graph Drawing*, 2004, pp. 206–216.

[77] V. Dujmovic and S. Whitesides, "An Efficient Fixed Parameter Tractable Algorithm for 1-Sided Crossing Minimization," in *Graph Drawing*, 2002, pp. 118–129.

[78] W. Barth, M. Jünger, and P. Mutzel, "Simple and efficient bilayer cross counting," in *Graph Drawing*, 2002, pp. 130–141.

[79] T. Eschbach, W. Günther, R. Drechsler, and B. Becker, "Crossing Reduction by Windows Optimization," in *Graph Drawing*, 2002, pp. 285–294.

[80] C. Gutwenger and P. Mutzel, "An Experimental Study of Crossing Minimization Heuristics," in *Graph Drawing*, 2003, pp. 13–24.

[81] P. Eades, "A heuristic for graph drawing," in *Congressus Numerantium*, 1984, pp. 149–160.

[82] P. Mutton and P. Rodgers, "Demonstration of a Preprocessor for the Spring Embedder," in *Graph Drawing*, 2002, pp. 374–375.

[83] T. M. J. Fruchterman and E. M. Reingold, "Graph drawing by force-directed placement," *Software, Practice and Experience*, vol. 21, no. 11, pp. 1129–1164, 1991.

[84] T. Kamada and S. Kawai, "An algorithm for drawing general undirected graphs," *Information Processing Letters*, vol. 31, pp. 7–15, 1989.

[85] A. Frick, A. Ludwig, and H. Mehldau, "A fast adaptive layout algorithm for undirected graphs," in *Graph Drawing*, 1994, pp. 389–403.

[86] P. Gajer, M. T. Goodrich, and S. G. Kobourov, "A multi-dimensional approach to force-directed layouts of large graphs," in *Graph Drawing*, 2000, pp. 211–221.

[87] J.-H. Chuang, C.-C. Lin, and H.-C. Yen, "Drawing Graphs with Nonuniform Nodes Using Potential Fields," in *Graph Drawing*, 2003, pp. 460–465.

[88] S. Hachul and M. Jünger, "Drawing Large Graphs with a Potential-Field-Based Multilevel Algorithm," in *Graph Drawing*, 2004, pp. 285–295.

[89] E. R. Gansner, Y. Koren, and S. North, "Graph Drawing by Stress Majorization," in *Graph Drawing*, 2004, pp. 239–250.

[90] E. D. Giacomo, W. Didimo, G. Liotta, and H. Meijer, "Computing Radial Drawings on the Minimum Number of Circles," in *Graph Drawing*, 2004, pp. 251–261.

[91] J. M. Six and I. G. Tollis, "A Framework for Circular Drawings of Networks," in *Graph Drawing*, 1999, pp. 107–116.

[92] J. M. Six and I. Y. G. Tollis, "A Framework for User-Grouped Circular Drawings," in *Graph Drawing*, 2003, pp. 135–146.

[93] M. Kaufmann and R. Wiese, "Maintaining the Mental Map for Circular Drawings," in *Graph Drawing*, 2002, pp. 12–22.

[94] M. Y. Becker and I. Rojas, "A graph layout algorithm for drawing metabolic pathways," *Bioinformatics*, vol. 17, pp. 461–467, 2001.

[95] P. D. Karp and S. Paley, "Automated Drawing of Metabolic Pathways," in *Third International Conference on Bioinformatics and Genome Research*, L. Hunter, D. Searls, and J. Shavlik, Eds. AAAI Press, 1994, pp. 207–215.

[96] G. D. Battista, P. Eades, R. Tamassia, and I. G. Tollis, "Algorithms for Drawing Graphs: an Annotated Bibliography," *Computational Geometry - Theory and Applications*, vol. 4(5), pp. 235–282, 1994.

[97] A. Liebers, "Planarizing Graphs - A Survey and Annotated Bibliography," *Journal of Graph Algorithms and Applications*, vol. 5, no. 1, pp. 1–74, 2001. [Online]. Available: http://www.cs.brown.edu/publications/jgaa/

[98] R. Diestel, *Graph Theory*. Springer-Verlag, 2005.

[99] M. Jünger and P. Mutzel, Eds., *Graph Drawing Software*. Berlin, Heidelberg: Springer-Verlag, 2003.

[100] J. Hopcroft and R. E. Tarjan, "Efficient Planarity Testing," *ACM*, vol. 21, no. 4, pp. 549–568, 1974.

[101] A. Lempel, S. Even, and I. Cederbaum, "An Algorithm for Planarity Testing of Graphs," in *Theory of Graphs: International Symposium (Rome 1966)*. New York: Gordon and Breach, 1967.

[102] S. Even and R. E. Tarjan, "Computing an st-Numbering," *Theoretical Computer Science*, vol. 2, 1976.

[103] K. Booth and G. Lueker, "Testing for the Consecutive Ones Property Interval Graphs and Graph Planarity Using PQ-Trees," *Journal of Computer and System Science*, vol. 13, pp. 335–379, 1976.

[104] N. Chiba and S. Nakano, "A Linear Algorithm for Embedding Planar Graphs Using PQ-Trees," *Journal of Computer and System Science*, vol. 30, pp. 54–76, 1985.

[105] G. D. Battista and R. Tamassia, "Incremental Planarity Testing," in *30th IEEE Symposium on Foundations of the Computer Science*, 1992, pp. 261–270.

[106] M. R. Garey and S. D. Johnson, "Crossing number is NP-complete," *SIAM Journal of Algebraic Discrete Methods*, vol. 4, pp. 312–316, 1983.

[107] C. F. X. de Mendonca Neto, "A Layout System for Information System Diagrams," no. 94-01, 1994.

[108] F. Schacherer, "A review of biological network visualisation," in *Pacific Symposium on Bioinformatics*, 2001.

[109] BioCarta, http://www.biocarta.com, 2002.

[110] R. D. Appel, A. Bairoch, and D. F. Hochstrasser, "A new generation of information retrieval tools for biologists: the example of the ExPASy WWW server," *Trends in biochemical Sciences*, vol. 19, pp. 258–260, 1994.

[111] ExPASY - Biochemical Pathways, http://www.expasy.ch/cgi-bin/search-biochem-index, 2005.

[112] M. Kanehisa and S. Goto, "KEGG: Kyoto Encyclopaedia of Genes and Genomes," *Nucleic Acid Res.*, vol. 28, pp. 27–30, 2000.

[113] KEGG, http://www.genome.ad.jp/kegg/pathway.html, 2005.

[114] V. J. Carey, J. Gentry, E. Whalen, and R. Gentleman, "Network structures and algorithms in bioconducter," *Bioinformatics*, vol. 21, pp. 135–136, 2005.

[115] Graphviz, http://www.graphviz.org/, 2005.

[116] BioMiner, http://www.zbi.uni-saarland.de/chair/projects/BioMiner/-index.shtml, 2005.

[117] P. Shannon, A. Markiel, O. Ozier, N. S. Baliga, J. T. Wang, D. Ramage, N. Amin, B. Schwikowski, and T. Ideker, "Cytoscape: A Software Environment for Integrated Models of Biomolecular Interaction Networks," *Genome Research*, vol. 13, pp. 2498–2504, 2003.

[118] B.-J. Breitkreutz, C. Stark, and M. Tyers, "Osprey: a network visualization system," *Genome Biology*, vol. 4, no. R22, 2003.

[119] J. A. Dickerson, D. Berleant, Z. Cox, W. Qi, and E. Wurtele, *Creating Metabolic Network Models using Text Mining and Expert Knowledge*. World Scientific Publishig Singapore, 2003.

[120] L. Krishnamurthy, J. Nadeau, G. Ozsoyoglu, M. Ozsoyoglu, G. Schaeffer, M. Tassan, and W. Xu, "Pathway database system: an integrated system for biological pathways," *Bioinformatics*, vol. 19, pp. 930–937, 2002.

[121] R. Pandey, R. K. Guru, and D. W. Mount, "Pathway miner: extracting gene association networks from molecular pathways for predicting the biological significance of gene expression microarray data," *Bioinformatics*, vol. 20, no. 13, pp. 2156–2158, 2004.

[122] A. Nikitin, S. Egorov, N. Daraselia, and I. Mazo, "Pathway studio - the analysis and navigation of molecular networks," *Bioinformatics*, vol. 19, pp. 1–3, 2003.

[123] E. Demir, O. Babur, U. Dogrusoz, A. G. amd G. Nisanc, R. Cetin-Atalay, and M. Ozturk, "PATIKA: an integrated pathway database with advanced visualization and querying facilities," *Bioinformatics*, vol. 18, no. 7, pp. 996–1003, 2002.

[124] A. Ludemann, D. Weicht, J. Selbig, and J. Kopka, "PaVESy: Pathway Visualization and Editing System," *Bioinformatics*, vol. 20, no. 16, pp. 2841–2844, 2004.

[125] PaVESy, http://pavesy.mpimp-golm.mpg.de/PaVESy.htm, 2005.

[126] Z. Hu, J. Mellor, J. Wu, and C. DeLisi, "VisAnt: an online visualization and analysis tool for biological interaction data," *BMC Bioinformatics*, 2004.

[127] M. Holford, N. Li, P. Nadkarni, and H. Zhao, "VitaPad: visualization tools for the analysis of pathway data," *Bioinformatics*, vol. 21, no. 8, pp. 1596–1602, 2005.

[128] K. Han, B. Hyon, and H. Jung, "WebInterViewer: visualizing and analyzing molecular interaction networks," *Nucleic Acids Research*, vol. 32, pp. W89–W95, 2004.

[129] YWays, http://www-pr.informatik.uni-tuebingen.de/~eiglsper/pathways/, 2002.

[130] F. Schreiber, "High quality visualization of biochemical pathways in BioPath," *In Silico Biology*, vol. 6, 2002.

[131] BioCyc, http://www.biocyc.org, 2004.

[132] N. R. Quinn and M. A. Breuer, "A force directed component placement procedure for printed circuit boards," *IEEE Transactions on Circuits Systems*, vol. 11, pp. 109–125, 1979.

[133] F. J. Brandenburg, M. Forster, A. Pick, M. Raitner, and F. Schreiber, "BioPath - Visualization of Biochemical Pathways," in *German Conference on Bioinformatics; Braunschweig*, E. Wingender, R. Hofestaedt, and I. Liebich, Eds., 2001, pp. 11–15.

[134] I. Rojdestvenski, "Objective and subjectiv relations in data visualization. examples from molecular biological data collections," in *Information Visualization*, E. Banissi, K. Borner, C. Chen, G. Clapworthy, C. Maple, A. Lobben, C. Moore, J. Roberts, A. Ursyn, and J. Zhang, Eds., 2003, pp. 544–548.

[135] YFiles, http://www.yworks.com/en/products_yfiles_about.htm, 2005.

[136] L. Stryer, *Biochemistry*. Heidelberg: Spektrum Akademischer Verlag, 1995.

[137] B. Oesterreich, *Objektorientierte Softwareentwicklung*. München, Wien: Oldenbourg Verlag, 1998.

[138] J. A. Robinson and T. P. Flores, "Novel techniques for visualizing biological information," in *5th International Conference on Intelligent Systems for Molecular Biology*. AAAI Press, 1997, pp. 241–249.

[139] A. M. MacEachren, *How Maps Work*. New York, USA: The Guilford Press, 1995.

[140] W. Müller and H. Schumann, "Visualization Methods for Time-Dependet Data - an Overview," in *Winter Simulation Conference*, S. Chick, P. J. Sanchez, and D. J. Morrice, Eds., 2003.

[141] M. C. F. de Olveira and H. Levkowitz, "From visual data exploration to visual data mining: a survey," *IEEE Transactions on Visualization and Computer Graphics*, vol. 9, no. 3, pp. 378–394, 2003.

[142] K. W. Brodlie, L. Carpenter, R. A. Earnshaw, J. R. Gallop, R. J. Hubbold, A. M. Munford, C. D. Osland, and P. Quarendon, *Scientific visualization.* Springer-Verlag, 1992.

[143] J. J. van Wijk and E. van Selow, "Cluster and Calendar- based Visualization of Time Series Data," in *IEEE Symposium on Information Visualization (InfoVis 99)*, G. Wills and D. Keim, Eds. IEEE Computer Society, 1999, pp. 4–9.

[144] F. Brian and J. Pritchard, "Visualisation of historical events using Lexis pencils, in: Case Studies of Visualization in the Social Sciences [online]," http://www. agocg.ac.uk/reports/visual/casestud/ contents.htm, 1997.

[145] C. Tominski, J. Abello, and H. Schumann, "Axes-based visualizations with radial layouts," in *Symposium on Applied Computing*, 2004, pp. 1242 – 1247.

[146] P. Dragicevic and S. Huot, "Spiraclock: a continuous and non-intrusive display for upcoming events," in *Conference on Human Factors in Computing Systems archive*, 2002, pp. 604 – 605.

[147] J. V. Carlis and J. Konstan, "Interactive visualization of serial periodic data," in *11th annual ACM symposium on User interface software and technology*, 1998, pp. 29–38.

[148] M. Weber, M. Alexa, and W. Müller, "Visualizing Time-Series on Spirals," in *IEEE Symposium on Information Visualization (InfoVis 01)*, 2001, pp. 7–13.

[149] J. Lin, E. J. Keogh, S. Lonardi, J. P. Lankford, and D. M. Nystrom, "VizTree: a Tool for Visually Mining and Monitoring Massive Time Series Databases," *VLDB*, pp. 1269–1272, 2004.

[150] L. Treinish and D. Silver, "Worm Plots," *IEEE Computer Graphics and Applications*, vol. 17, pp. 17–20, 1997.

[151] T. Dwyer, H. Rolletscheck, and F. Schreiber, "Representing experimental biological data in metabolic networks," in *Proceedings of the second conference on Asia-Pacific Bioinformatics*, vol. 29, 2004, pp. 13 – 20.

[152] M. Himsolt, http://www.infosun.fmi.uni-passau.de/Graphlet/download/-misc/GML.ps.gz, 2005.

[153] C. Johnson, "Top Scientific Visualization Research Problems," *IEEE Computer Graphics and Applications*, vol. 24, no. 4, pp. 13–17, 2004.

[154] B. Shneiderman, "The Eyes Have It: A Task by Data Type Taxonomy for Information Visualizations," in *IEEE Symposium on Visual Languages table of contents*. Washington, DC, USA: IEEE Computer Society, 1996.

[155] D. A. Carr, "Guidelines for designing information visualization applications," in *ECUE'99*, 1999.

[156] H. Schumann and W. Müller, *Visualisierung*. Berlin: Springer-Verlag, 2000.

[157] B. Shneiderman, *Designing the user interface*. Addison-Wesley Publishing Company Inc, 1992.

[158] T. Dwyer and P. Eades, "Visualising a Fund Manager Flow Graph with Columns and Worms," in *Proceedings of IEEE IV'02*. IEEE Computer Society, 2002, pp. 147–158.

[159] G. K. Batchelor, *An introduction to Fluid Dynamics*. Cambridge: Cambridge University Press, 1967.

[160] M. Grand, *Patterns in Java: a catalog of reusable design patterns*. New York: Wiley Computer Publishing, 1998.

[161] OpenGL Architecture Review Board, D. Shreiner, M. Woo, J. Neider, and T. Davis, *The OpenGL Programming Guide - The Redbook*, 4th ed. Addison-Wesley Pub Co, 2003.

[162] JOGL, https://jogl.dev.java.net/, 2005.

[163] P. Kuchana, *Software architecture design patterns in Java*. Auerbach Publications, 2004.

[164] H. Rinne, *Statistische Analyse multivarianter Daten.* Wien-Oldenburg Verlag, 2000.

[165] A. L. Edwards, *An Introduction to Linear Regression and Correlation.* San Francisco: W. H. Freeman and Company, 1976.

[166] P. Y. Chen and P. M. Popovich, *Correlation: Parametric and Nonparametric Measures.* Thousand Oaks, CA: Sage University Paper Series on Quantitative Applications in the Social Sciences: 07-139, 2002.

[167] K. Backhaus, B. Erichson, W. Plinke, and R. Weiber, *Multivariate Analysemethoden.* Berlin, Heidelberg, New York: Spriger-Verlag, 2003.

[168] K. Freitag, *Zeitreihenanalyse.* Köln: Lohmar Verlag, 2003.

[169] H. Kantz and T. Schreiber, *Nonlinear time series analysis.* Cambridge: Cambridge University Press, 2004.

[170] P. Diggle, *Time Series.* New York: Oxford Science Publications, Oxford University Press, 1995.

[171] L. Laloux, P. Cizeau, J.-P. Bouchaud, and M. Potters, "Noise Dressing of Financial Correlation Matrices," *Physical Review Letter*, vol. 83, pp. 1467–1470, 1999.

[172] V. Plerou, P. Gopikrishnan, B. Rosenow, L. A. N. Amarel, and H. E. Stanley, "Universal and Nonuniversal Properties of Cross Correlations in Financial Time Series," *Physical Review Letters*, vol. 83, p. 1471, 1999.

[173] V. Plerou, P. Gopikrishnan, B. Rosenow, L. A. Amaral, T. Guhr, and H. E. Stanley, "Random matrix approach to cross correlations in financial data," *Physical Reviews E*, vol. 65, p. 066126, 2002.

[174] M. Mueller and G. Baier, "Detection and characterization of the correlation structure in multivariate time series," *Physical Reviews E*, vol. 71, p. 046116, 2005.

[175] A. Pikovsky, M. Rosenblum, and J. Kurths, *Synchronization.* Cambridge: Cambridge University Press, 2001.

[176] M. G. Rosenblum, A. S. Pikovsky, and J. Kurths, "Synchronization approach to analysis of biological systems," *Fluctuation and Noise Letters*, vol. 4, no. 1, pp. L53–L62, 2004.

[177] M. J. Berridge, M. D. Bootman, and P. Lipp, "Calcium - a life and death signal," *Nature*, vol. 395, pp. 645–648, 1998.

[178] A. Z. Larsen and U. Kummer, *Understanding Calcium Dynamics*. Lecture Notes in Physics, 2003.

[179] N. M. Woods, K. S. R. Cuthbertson, and P. H. Cobbold, "Repetitive transient rises in cytoplasmic free calcium in hormone-stimulated hepatocytes," *Nature*, vol. 319, pp. 600–602, 1986.

[180] S. Schuster, M. Marhl, and T. Hofer, "Modelling of simple and complex calcium oscillations: From single-cell responses to intercellular signalling," *European Journal of Biochemistry*, vol. 269, no. 5, pp. 1333–55, 2002.

[181] U. Kummer, L. F. Olsen, C. J. Dixon, A. K. Green, E. Bornberg-Bauer, and G. Baier, "Switching from simple to complex oscillations in calcium signaling," *Biophysical Journal*, vol. 79, no. 3, pp. 1188–1195, 2000.

[182] D. T. Gillespie, "Exact stochastic simulation of coupled chemical reactions," *Journal of Physical Chemistry*, vol. 81, no. 25, pp. 2340–2361, 1977.

[183] R. J. Muirhead, *Aspects of multivariante statistical theory*. New York: John Wiley and Sons, 1982.

[184] JAMA, http://math.nist.gov/javanumerics/jama/, 2005.

[185] Gnuplot, http://www.gnuplot.info/, 2005.

Glossary

3D: three dimensions, 50

BFS: **B**readth **F**irst **S**earch, 31

DFS: **D**epth **F**irst **S**earch, 30

ExPASY: **Ex**pert **P**rotein **A**nalysis **SY**stem, 41

GUI: **G**raphical **U**ser **I**nterface, 12

KEGG: **K**yoto **E**ncyclopedia of **G**enes and **G**enomes, 42

MVC: **M**odel-**V**iew-**C**ontroller, 25

PO: **P**eroxidase-**O**xidase reaction, 51

SBML: **S**ystems **B**iology **M**arkup **L**anguage, 18

SVG: **S**uper **V**ector **G**raphics, 22

ViPaSi: **Vi**sualisation of biochemical **P**athways and their **Si**mulation results, 14

Bibliography

190

Index